True Planetary Motions And Rhythmic Climatic Changes

True Planetary Motions And Rhythmic Climatic Changes

Manhin Look-Yat
&
Helen Look-Yat Taylor

To order additional copies of this book, contact:
Xlibris Corporation
1-888-795-4274
www.Xlibris.com
Orders@Xlibris.com
86175

Contents

FOREWARD

TRUE PLANETARY MOTIONS AND RHYTHMIC CLIMATIC CHANGES
By Manhin Look-Yat 1925 ~ 1996

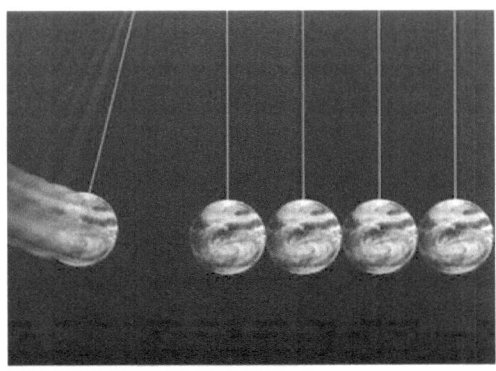

World climate has been swinging between ice-ages with precision. Many Climatologists think that the chief cause of these repeated swings is a change in the intensity of sunlight as a result of shifts in the tilt of the Earth's axis. These climate shifts in the past have caused major waves of extinction.

To this day, no one could find the answers or reasons for these changes, e.g. cyclical ice-ages, hot-spells, reversals of magnetic poles and mass extinctions. The answers are in the concept of an orbiting Sun. In my projection, the observer is placed outside our galaxy. From that vantage point he sees our Sun as a medium sized star doing a spiral circle around the inner hub of our galaxy. The rest follows: All the known planets are seen circling our Sun on a single plane in concentric circles, crossing the plane of the Sun's orbit at right angles at all times. **In the past, man thought the Sun was stationary, hence man never discovered true planetary motions!**

Attached is a manuscript on true planetary motions by Manhin Look-Yat, prepared over rigorous study and observation, concluded with the discovery of the century, edited and transcribed by his daughter Helen Look-Yat Taylor: **"TRUE PLANETARY MOTIONS & RHYTHMIC CLIMATIC CHANGES"**

DEDICATION

In memory and dedication to the legacy of my father's life work, I dedicate this book with eternal admiration, gratitude and respect to the world community on his behalf and to the annals of Science and Astronomy. The brilliance of this discovery and concentrated works revealed in this manuscript will change forever man's perception of our galaxy and the universe. I take comfort in knowing from his projections and discovery herein that the world will not "end" in 2012 as sensationalism predicts, but that our planet is simply climaxing into a new cyclical age. Many of the yet unanswered questions that have eluded mankind ever since he emerged from the last ice-age will now be answered within this manuscript.

At a point some thirteen years after his suicidal death from the ravages of Parkinson's disease, I stand witness to attest to the dedication and conviction with which he so painstakingly produced this manuscript, typing one finger at a time on a typewriter I gave to him in encouraging him to produce this work. May all the years and sacrifices he made for this his "baby", his life's toil and meditation be a living legacy in the annals of the discoveries of man. He noted in his writings on this life's work that he would hope it would not take another five thousand years after mankind creeps out of the next ice-age to un-earth this important discovery! He was concerned about taking this immense knowledge with him to his grave as he worked feverishly to complete it before his passing in 1996. I vowed then upon reading this to make this labour of love to his legacy in making this manuscript known to the world. Upon grasping the revelations of his astronomical projections on true planetary motions, one will find some astonishing similarities to the ancient Maya in their projections of cyclical life cycles. Heeding the warnings and timely projections of his discovery will prove to be a crucial guide within the next 1,000 years toward survival of mankind and all living things through the next life cycles. The jigsaw puzzle pieces will now all fit as mankind seeks still to unravel unanswered mysteries of our universe.

For my dear mother, Ernestine, his wife of 50 years upon his passing, eternal gratitude, love and undying devotion, as you gave to him during his life. For my sisters, Linda, Jacqueline, Ingrid, Corinne and Jennifer~ I take the baton in completion of this book in heralding the love, respect and gratitude we all feel for him forever.

Helen Look-Yat Taylor

PREFACE

True planetary motions are no longer a mystery. In this paper, the reader knows from the start that the butler did it. The key to my discovery lies in the concept of a moving Sun. We now know that our galaxy is spiral and that the Sun is a medium sized star orbiting the inner hub of our galaxy, so we found the key.

In this projection I discovered true planetary motions by placing myself as an observer at points outside our galaxy. The reader must do the same. From that point I made the following observations:-

I saw billions of stars moving in spiral formation to form this Milky Way galaxy. I saw our Sun as a medium sized star circling the inner hub of the galaxy. I saw the nine known planets moving around our Sun in concentric circles, all on the same plane, their orbits always crossing the plane of the Sun's orbit at right angles.

I knew then that I discovered true planetary motions and that Kepler was wrong and that Einstein never dreamed of this. I saw the Earth's axis pointing in the direction of Polaris when the Sun was at 6:30, 5:30, 11:30, and 12:30 points on its orbital plane. The visualization of the Sun's orbit is from a reference to the points on a clock as outlined above. Then it pointed to Thurban and Vega as the Sun went on to point 9 and point 3. I saw Polaris, Earth's pole star, 57 light years away from the Sun, with Earth's axis pointing directly at Polaris at this point in time in 1994. Since there is evidence that the last ice-age on Earth was 11,000 years ago, I arrived at the conclusion that the Sun makes each orbit in 25,000 years precisely; and that they, ice-ages, occur simultaneously in both hemispheres. The timing is precise!

I saw the axis of the planets pointing in fixed directions relative to other stars in the galaxy. I extended the orbital planes of all the known planets in a concentric way and observed that each planetary orbit when extending, pierces an imaginary point at the center of the Sun's orbital plane. I observed that as the Sun slowly changed its position on its orbit along the hub of the galaxy, that the planetary orbits continue to cross the plane of the Sun's orbit at right angles at all times bringing about catastrophic climatic changes on Earth and other planets as the Sun made its rounds cyclically. **I knew then I found that the laws had to be changed.**

I went on to draw up these laws:-

THE LAWS OF PLANETARY MOTIONS by Manhin Look-Yat, 1994

(1) The Sun must move in a spiral circle on the inner hub of the galaxy once every 25,000 years, precisely.

(2) All nine known planets, plus the two which were newly discovered by the Chinese and named Purple Mountain 1 and 2, must orbit the Sun on the same plane at all times forming concentric circles which when extended must intersect with the center of the Sun's orbit while crossing the plane of the Sun's orbit at right angles at all times.

(3) There shall be no wobble or gyration or nutation of Earth's axis or of the axes of other planets in the galaxy. **ALL** axes of the planets **MUST** point in fixed directions relative to other stars as the Sun makes its rounds.

KEPLER WAS WRONG. <u>He saw the Sun in a fixed state. This discovery eluded man for all time because man saw the Sun stationary.</u>

Nothing is left to chance. There is complete order in the universe. Seasonal and catastrophic changes occur on Earth and other planets only because the Sun moves in spiral circles. This causes distribution of sunlight on the planets, as the Sun changes its position on its orbit.

Note:

The Earth and other planets do not have the ability to steer their axes in any direction as the Sun curves in space. Since their axes point in fixed directions relative to the stars, sunlight shifts. This causes catastrophic climatic changes to occur every 25,000 years. In the case of the Earth, prevailing conditions of seasonal changes occur only when the Sun is at point 5:30, 6:30, 11:30 and 12:30 on its orbital plane.

ICE-AGES

Ice-ages come on when the Sun gets to point 12 and point 6 on its orbital plane using the point of reference as outlined visualizing the face of a clock. **The Sun moves in an anti-clockwise direction on its orbital plane.** At points 12 and 6 there are two ice-ages with each point completed by the Sun.

The last ice-age was 11,000 years ago. **We are already into the present ice-age!** It will last another 2,000 years. This discovery unravels mysteries of shifting magnetic poles, ice-ages, catastrophic climatic changes, cyclical population explosions, extinction of species, rising oceans, etc.

This is truly the dawning of a new day for man.

Manhin Look-Yat
August 6th, 1925~ July 1st, 1996

TRUE PLANETARY MOTIONS AND RHYTHMIC
CLIMATIC CHANGES

By MANHIN LOOK-YAT
&
HELEN LOOK-YAT-TAYLOR

ABSTRACT

This paper on our shifting Sun reports the discovery of true planetary motions which makes possible long-term weather forecasting of ice-ages, hot-spells and magnetic pole shifts on Earth. It throws new light on the green house effect. By this discovery, the Sun is now known to be moving in spiral circles on the inner hub of our galaxy. **Observation of the moving Sun has been ignored but it reveals the key to true planetary motions.**

<u>In future, all planetary motions must be made from an imaginary point at the center of the orbit traced by the moving Sun as it circles the inner hub of our galaxy as seen by an observer outside our galaxy.</u>

Astronomers from THALES and PYTHAGARUS to PTOLEMY, believed that the Earth was the center of the universe. That theory held for 1,400 years until COPERNICUS(1,473-1,543), demonstrated that the Sun was stationary at the center of the universe. He too was wrong, yet we also accepted his theory for 400 more years to this day.

This discovery of a moving Sun is the key to true planetary motions. The Sun is now known to be moving anti-clockwise. The planets can be seen moving with the Sun on a single plane in concentric circles. There is complete harmony in their movements. Their orbits cross the Sun's orbit at right angles at all times in a concentric way while the Sun does a spiral and that's because the galaxies also move in spirals.

The objectives of this paper are:-

To show a direct link between the movement of the Sun, true planetary motions, cyclical reversals and periodical fading of the magnetic poles of the Earth. This makes possible long term forecasting of the global green house effect and climatic changes. In addition it will enable us to harness and control population growth and population explosions that occur four times every 25,000 years as the Sun makes its round on the inner hub of our galaxy. (Not 250 million years as previously believed).

It shows that there is complete harmony and order in the universe. No gyration, no wobbles, no nutation, of planets. The orbiting Sun holds the key to true planetary motions and the so called green house effects occurring on Earth today. This is truly the dawn of a new day. Future forecasting is now possible. Cyclical global warming is now predictable because of this observation.

We are observing changes in climate which we do not clearly understand. We shall still focus on the green house effect and pollution, but the effect of those are too infinitesimal to be compared to the effects on climate caused by the moving Sun. Our attention should now shift to planetary motions in order to halt the population explosion now in progress. World population is 5.2 billion. This figure will double in 30 years. **WE MUST ACT NOW.**

INTRODUCTION

Ever since man became curious and looked towards the stars his intelligent enquiring mind began to ask questions and to seek the answers. Today man still has a problem with long term forecasting. Modern man looks back on barely 5,000 years of recorded history. This is so mainly because of devastating cyclical climatic changes that over-come the Earth every 12,500 years. E.g. ice-ages and alternating hot-spells, as the Sun orbits the inner hub of our galaxy every 25,000 years. We know of these changes but we have no record of these devastating changes that occur slowly as the Sun curves in space.

Today man is still erroneously making his projection of true planetary motions from the point at the center of the Sun itself, with no consideration for the Sun's movement on the hub of this galaxy. We perceive the Sun as stationary. We are still hanging on to KEPLER's theory on planetary motion, which is still wrong. To date, Kepler's theory is still accepted. He was wrong. Kepler saw the Sun as stationary as he looked at it from Earth.

The key to this solution always eluded man. This projection will now be done from a position of considering a moving Sun. The observer is placed at a point at the center of the Sun's orbit (an imaginary point), not the center of the stationary Sun. From this vantage point the observer sees all planetary orbits when extended in a concentric way piercing the center of the circle traced by the Sun, while crossing the orbital plane of Sun at right angles at all times. This discovery is the answer to a multitude of questions. All the planets will be seen on a single plane with their axes pointed in fixed directions relative to other stars in the galaxies.

Questions relative to true planetary motions, precise timing of ice-ages, prediction of long term climatic changes, cyclical reversal of the magnetic poles, the extinction of the dinosaurs, the occurrence of cyclical hot-spells every 12,000 years between ice-ages, rising oceans, cyclical population explosions etc. The list is endless. Population explosion or growth must be harvested immediately.

We are focusing on tectonics or shifting continental plates, instead of observing the Sun that does a spiral orbit on the inner hub of our galaxy and that is wrong.

We do not have to move the continents on plates as thin as toilet paper relative to the thickness of the Earth's crust, to account for the tropical coral reefs that are today located in places like Spitsbergen in the Arctic and other Antarctic regions of the globe. The discovery of an orbiting Sun is a brilliant milestone in the history of astronomy. **This is a breakthrough! We can now forecast long-term cyclical climatic changes that occur with precision timing over thousands of years.**

At this point in time, man can only do forecasts of three to four days at a time (outside of seasonal changes). This projection makes it brilliantly clear that all climatic changes on Earth or on all other planets on our solar system are geared to the true movements of our Sun, which can only be observed from outside our galaxy. At last man can observe the Sun doing a spiral. He no longer sees it sitting there with the planets going around it. Unfortunately, Kepler and all before him, saw the Sun in a fixed state. In this projection, the results are fantastic. All planets are seen on the same plane—crossing the plane of the Sun's orbit at right angles at all times. In this formation, none of the planets can ever over-take the Sun as the Sun makes its round on the inner hub of our galaxy, while the Earth's axis points in the general direction of Polaris or Vega or Thurban.

Today Polaris is the North star or pole star, located 57 light years away and direct north ONLY because the Sun is at its present location on its orbit. As the Sun makes its round it describes a colossal spiral circle which is so large that it appears to be a straight line wherever the observer locates the Sun on it. In the past, it was universally agreed that the stars and planets were fixed to crystal spheres centered on the Earth. Then the idea of motions in empty space blossomed, followed by the idea of modern cosmological thought. This is a new and more widely embracing paradigm. My discovery of true planetary motions has come to replace the others.

This is the final solution to long-term forecasting. *THE DAWN OF A NEW DAY. THE DISCOVERY OF THE CENTURY* and of all time, this projection makes never before long-term forecasting possible.

The drawing shows Earth's axis today off the plane of the Sun's orbital plane by exactly 23& ½ degrees.

Earth, the "Blue Planet"

The axis appears to tilt, but it doesn't. The degree of tilt of the Earth's axis relative to the Sun's orbit is determined by the location of the Sun on its orbit. This information allows us to pinpoint the location of the Sun on its orbital plane which is numbered from 1-12 as on a clock. There is evidence that the Sun is moving anti-clockwise. There is also evidence that the last hot-spell on Earth was some 5,000 to 6,000 years ago. The Sun was then at point 9 on its orbit. The apparent 23 and ½ degree tilt of Earth's axis relative to the Sun's orbit occurs only when the Sun is either at one of these points on its orbit: At 6:30, 5:30, 11:30 or 12:30.

The degree of tilt of Earth's axis changes with each new position of the Sun on its orbit, yet the axis remains in a fixed direction relative to other stars in the galaxy. Today the Sun is moving into the 6:00 or 12:00 point where the ice-age occurs simultaneously in both hemispheres on this our Earth. This means that we are already half way into an ice-age reaching the high point in the next 1,000 years. At point 6 or 12, the degree of tilt of Earth's axis and Sun's orbit is reduced to zero degrees and Earth's axis and Sun's orbital plane will be parallel. **We can now forecast ice-ages, we can anticipate rising oceans and population explosions. We can now make plans to harness population growth. This would alleviate untold suffering brought about by population explosions between ice-ages.**

(See drawings, showing diagrams of the positioning of the Sun at points on clock representing the ice-ages)

Man has no control over the creeping ice belts but we must adjust. This ice-age will last for three thousand years. **Few of us will survive only if we live in the equatorial belt.**

Ice-ages occur only when the Sun is at point 6 or at point 12 on its orbit. Prevailing conditions occur when the Sun moves to any one of four points on its orbit, 5:30, 6:30, 11:30 or 12:30. When the Sun gets to point 3o'clock the Earth's axis lies in the plane of its own orbit and its axis is 90 degrees off the orbital plane of the Sun. At point three—the next hot-spells come. There are six months of darkness and six months of daylight each year as the Earth goes around the Sun. Its orbit then is in line with its own axis. The ice-age comes next as the Sun goes to point 12. These are the worst periods for all living things on Earth. During the ice-age, nearly all living things outside the equatorial belts die.

Population Explosions are Cyclical.

As we approach the next ice-age, man is unable to cope with disease, hunger, filth that will grow around him ten-fold. He too will die as the ice-belts close to within 10 or 15 degrees from the equator. It is already happening in the streets of Calcutta in India as we approach the ice-age. It can also be seen in the streets of California and Somalia . . . the population explosion is on, it precedes every ice-age and every hot-spell.

This projection should make us aware of the devastating and uncontrollable catastrophes that overcome Earth, one of which is moving in on us today.

China is now leading the way in harnessing population growth. Today we live in one of the best periods. The Sun is now at 6:30 point on its orbit, in 1,000 years the Sun will be at point 6, and it is another ice-age in both hemispheres simultaneously. We are well into an ice-age today. <u>We are halfway there!</u>

The Green House Effect and Population Explosions

The extinction of the dinosaurs is also linked to the moving Sun. Their demise was a direct result of land separation brought about by rising oceans between ice-ages. Cyclical ice-ages occur every 12,000 years. There is evidence that the last ice-age ended 11,000 years ago when our Sun moved away from point 12 on its orbit.

The theory of continental drift or tectonics is also reduction and absurdum.

With this projection, we are better prepared to deal with cyclical droughts, hot-spells and other catastrophic climatic changes that occur over and over in geologic time. Since we know that these periods are cyclical and directly linked to the moving Sun, they are now predictable in a precise way. Forecasting has arrived.

We must now divert our attention from the green house effect as we see it today. The green house effect is real, but for other reasons given by N.A.S.A and others.

We must find a way immediately to control population growth. There is no word to describe the magnitude of human suffering that precedes the ice-age already half-way in both northern and southern hemispheres, lasting three thousand years. We have to deal with the problem of too many mouths to feed beginning in the next one hundred years.

In the next one hundred years the population will be 30 billion, that is six times what it is today! Over population is occurring as the ice-age creeps up on us. In 1,000 years it becomes uncontrollable, then it is death and misery all around us for 3,000 years. We must be prepared. It is almost too late, the ice-belts are slowly closing in on the equator—(believe it or not).

Cyclical Ice-ages and Extinctions

These cycles occur twice as our Sun makes each orbit. In other words, there is an ice-age when the Sun gets to point six (6) on its orbit and again at point 12 on its orbit. This means that it is a 12,000 year rhythm, the timing is precise. World population went over the five billionth mark in 1987, in 20 years, it doubles, by the year 2030 it would be 30 billion. The terrible catastrophe has begun.

After the next ice-age, the Sun moves on to the 3 o'clock point on its orbit. At that point 6,000 years from now, another hot-spell occurs. The last hot-spell period ended 5,000 years ago, when the Sun moved from point 9. Peking man and the Neanderthals didn't quite make it, we must not let it happen again. We are at present in a runaway population growth situation and just beginning to be aware of it.

Great civilizations of the Americas vanished during the last ice-age 11,000 years ago because they didn't move with the ice-belts. The rising oceans between ice-ages destroyed the last land bridges between the continents 60 million years ago so that when the hot-spells came between ice-ages, the Dinosaurs were finally trapped in areas of total darkness for six months each year, then, not being able to find their way to sunlit areas, gradually starved to death. The destruction of these land bridges 60 million years ago brought about their final extinction. The process of extinction was very slow. Every ice-age brought them closer and closer to the end, until the last land bridges were gone.

The Logical Conclusion

It is time, now in knowing these True Laws of Planetary Motion that we must accept the fact that the Sun moves in spiral circles around the inner hub of our galaxy. The Sun moves in spiral circles and the planets move along with it, entrapped by the Sun. **Therefore all planetary motions must be made in consideration of a moving Sun. We have no alternative.**

There is a Nobel Prize out there for anyone who discovers true planetary motions. I deduced three (3) Laws on True Planetary Motions which will be backed up by drawings showing the behavior of the planets as the Sun moves from point to point on the inner hub of our galaxy. All points on the ellipse are numbered from 1 to 12 as on a clock, showing the Sun moving anti-clockwise on its cyclical orbit.

In Astronomy Made Simple, Meir H. Degani wrote that all objects in our galaxy rotate. He said the shape of the galaxy implies that it is rotating and that the stars rotate around the center of the galaxy in as much the same way as the planets move around our Sun with an orbital velocity of 160 miles per second. Since this is an accepted theory, the logical conclusion must follow and that is:—as long as the Sun follows an orbital plane around the hub of our galaxy, the climatic conditions on Earth and other planets in the solar system must be affected. The orbiting Sun produces shifting ice-belts and ice-ages on Earth with each round the Sun makes on its orbital plane this I can explain in detail.

This paper covers:

1. The Orbiting Sun Concept
2. Long Term Climatic Changes
3. Reversals of the Earth's Magnetic Poles
4. Ice-Ages, Glaciations and Hot-Spells
5. Continental Drift or Plates, Rising Oceans and Ocean Floors
6. Life on Other Planets
7. Extinction of the Species and Vanishing Civilizations
8. Gravitational Constant, Magnetic Orientation, Earth's Rotation
9. Droughts, Famine, etc
10. My three Laws of True Planetary Motions resulting in Rhythmic Climatic Manifestations

DISCOVERY OF THE CENTURY!

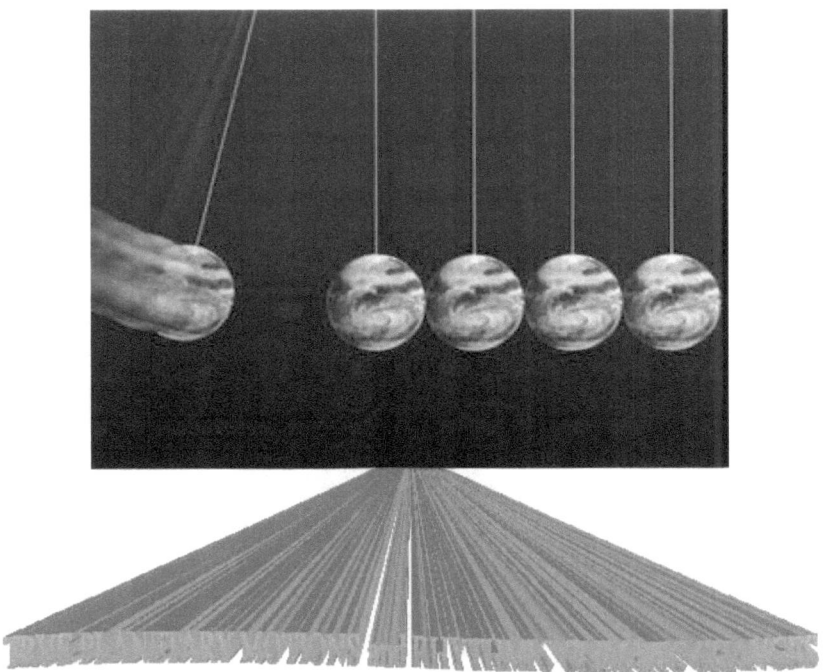

TRUE PLANETARY MOTIONS and
RHYTHMIC CLIMATIC CHANGES

By Manhin Look-Yat

Manhin Look-Yat 1925-1996

Co-authored by Helen Look-Yat Taylor

Chapter 1

GLOBAL LONG TERM FORECASTING
IS NOW MADE POSSIBLE!

As Above so Below

Planetary Motions is no longer a mystery. In this paper, the reader knows from the start *THAT THE BUTLER DID IT!* The key to my discovery lies in the concept of a moving Sun. We now know that our galaxy is spiral and the Sun is a medium sized star orbiting the inner hub of our galaxy. I discovered and drew up the following three laws of planetary motions. We have now found the key. It is the dawning of a new age for man. The day of long-term forecasting and prediction of catastrophic changes is here. We will now be able to predict accurately, and prepare for the cyclical rhythmic occurrence of ice-ages, glaciations, hot-spells and the accompanying population explosions, rising oceans and strategic migration for survival of these periods for millennia to come.

With my projections, the Sun orbits in its ellipse for 25,000 years in each orbit, with two ice-ages per cycle. The last ice-age was gauged at 11,000 years ago. On my discovery projection clock, an ice-age occurs when the Sun in its orbit on the inner hub of our galaxy, and arrives at points 6 and points 12 on the clock. Catastrophic climatic changes occur every 25,000 years. We are at point 6:30 in this cycle, as the Sun moves anti-clockwise in its orbit. We are already into the present ice-age, evident by the accompanying pre-existing climatic manifestations. **Earth is spiraling into the climax of an ice-age within 1,000 years, that will last another 2,000 years into the next hot-spell.**

The Discovery of the Century

<u>My Three Laws of Planetary Motions</u>

1. The Sun moves in spiral circles on the inner hub of our galaxy **once every 25,000 years**

2. The nine known planets plus the newly discovered by the Chinese, Purple Mountain 1, and Purple Mountain 2, **must orbit the Sun on the same plane, forming concentric circles while crossing the plane of the Sun's orbit at right angles at all times.**

3. There shall be no *"wobble"* or *gyration nor nutation* of Earth's axis or the axis of other planets in the galaxy. All axes of other planets in the galaxy **MUST** point in fixed directions relative to other stars as the Sun makes its rounds.

Chapter 2

A BRIEF HISTORY OF ASTRONOMY

The Geocentric Period

In the geocentric period, man thought he understood well the motions of the Earth's axis. Ptolemy and other early astronomers believed that the Earth was the center of the universe and that all other objects revolved around it. Then Copernicus in 1473-1543 demonstrated that the Sun was the center of the Universe, not the Earth. Although he was wrong, THAT THEORY WAS ACCEPTED WITHOUT QUESTIONING FOR 400 YEARS. Unfortunately, Copernicus too, saw the Sun in a fixed state.

The Galactic Period

In this period, Copernicus demonstrated that the Earth revolved around the Sun along with all other planets. This was something new. It came 1,400 years after Ptolemy's theory. In the year 1610, Galileo introduced the telescope, another milestone in the history of astronomy.

Kepler (1571-1630), came up with HIS three laws on planetary motions, which had not only short comings but were entirely wrong. To my amazement his laws are still in the books in 1993! Kepler never did explain why the planets behaved in the way he said they did. He was wrong because he made his projections from a point at the center of the Sun. He too saw the Sun as **stationary.**

The Universal Period

We are in that period today. It is now apparent that the galaxy of stars to which our Sun belongs is merely one of billions of galaxies. This is the period in which we live. Yet, brilliant minds like that of Milankovich, Einstein, Carl Sagan, et al—made the same mistake. NOT ONE OF THEM made their projections relative to a moving Sun. It never occurred to them that they should make their projections on planetary motions from the point at the center of the orbit formed by the Sun.

They failed to discover True Planetary Motions because they failed to observe that ALL planetary orbits when extended in a concentric way, touch the point at the center of the orbital plane of the Sun.

TRUE PLANETARY MOTIONS is the key to ***LONG TERM FORECASTING***.

In order to carry out this experiment the observer must be placed at a point outside the galaxy.

Early and Late Astronomers

The Chinese studied the stars as far back as 3,000 B.C. Our recorded history only goes back 5,000 years. Ptolemy, an Egyptian, lived within a period around 3,000 B.C. His theory that the Earth was at the center of the universe stood for about 1,400 years, until Copernicus. Copernicus 1473-1543 who demonstrated that the Earth went around the Sun and that the Sun was the center of the Universe and this was accepted for another 400 years ago, up to the last 20 years, Copernicus too was wrong.

They all made the same mistake. NOT ONE OF THEM made their projections relative to a moving Sun. It never occurred to them that they should make their projections on planetary motions from the point at the center of the orbit formed by the Sun. None, not one of them!

In the year 230 B.C. Eratosthenes made a catalog of 675 fixed stars. The first astronomer we know of was Thales, born in 640 B.C. and died in 556 B.C. As far back as 550 B.C.

Pythagarus thought that in fact the Earth was a globe and not a flat disc surrounded by water. He was right but no one believed him until Columbus proved it.

Hypparchus was Greek. He was born in the year 170 B.C. He was considered the greatest of all the ancient astronomers. Tycho Brahe 1546-1601 was born in Denmark. None of these men—from Thales to Carl Sagan, ever found the key to true planetary motions. *It simply never occurred to them that a spiraling Sun always influenced true planetary motions.*

Chapter 3

MODERN ASTRONOMY

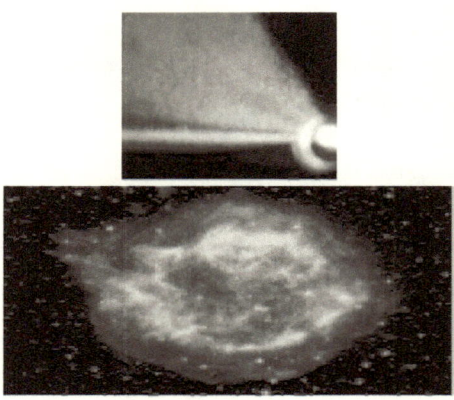

Modern Astronomy begins with the leading figures like Galileo and Kepler. Kepler was German, born in 1571. He came up with three laws on planetary motions, which, absurd as they seem to me today, stood for four hundred years to this day. **He was wrong for the same reasons. He too, used the stationary Sun as the center in his projection of his laws**.

In my projection on true planetary motions, the observer stands at a point <u>outside</u> our galaxy. Kepler too, saw the Sun in a fixed state. Galileo was from the same period. While Kepler was working in Germany, Galileo 1564-1642, born in Pizza, was experimenting in Italy. Others like William Herchel, John Elert, Borde, Le-Verier, Einstein, Carl Sagan, Milankovitch et-al, were thought to be brilliant, yet all of them made the same mistake. This idea eluded all of them. They never once used the point at the center of the Sun's orbit to make projections on planetary motions. They all used the Sun itself as the center.

THERE IS NO OTHER WAY TO DETERMINE TRUE PLANETARY MOTIONS BUT BY MOVING THE SUN IN SPIRAL CIRCLES ON THE INNER HUB OF OUR GALAXY.

The reason they never discovered true planetary motions is now clear. <u>They all thought that the Sun, being at the center of the solar system, stood still, and</u>

recently they came to believe that the Sun went along a straight course from a cloudy birth to a frozen extinction.

This is illustrated in Time Life books 1973 (drawing by Mel Hunter)
Time Life shows the Sun from a cloudy birth to a frozen extinction on a curved path in 50 billion years. **My discovery is a major break-through in forecasting. The pattern is precise, the timing is cyclical, long-term forecasting is here.**

It is 1993, and man still has a problem with forecasting on a long term basis, cyclical climatic changes and reversals and fading of the magnetic poles.

Observation of the moving Sun is the key to long-term forecasting. This observation has eluded man throughout history from Thales and Pythagarus, to Milankovitch, Einstein and Carl Sagan, et-al.

The Celestial Sphere

Let me explain the celestial sphere. The celestial sphere serves as a background upon which the stars are projected. It is concentric with the terrestrial sphere but infinitely larger. It is imaginary and no definite value is given to its radius.

For example: If the north pole of the Earth were extended infinitely, it would pierce the North star POLARIS at all times, as long as the Sun is at its present location at 6:30 on its orbit.

Let us look back 5,000 years. At that time the star THURBAN was our pole star, or North star, because the Sun was then at point 9 on its ellipse and the Earth's north axial pole when extended pierced Thurban. (as it pierces Polaris today). The Egyptians venerated it. Thurban was seen as the only stationary star as Polaris is today. As seen from the Earth all other stars described circles around it. At other times when the Sun was at point 12 the North star was either VEGA or again POLARIS.

The Sun takes 25,000 years to make one orbit on the inner hub of our galaxy and as the Earth continues to orbit the Sun wherever it goes, we find that because of this the Earth's axis describes a small circle embracing Polaris, Vega and Thurban.

Kepler's Laws

Here are Kepler's three laws on planetary motions. Kepler's Laws were based on the concept of a <u>stationary Sun.</u> Although Kepler never explained all the short comings of his three laws, they were accepted by the common herd and still stand after four hundred years.

<u>He was wrong because he could have never discovered true planetary motions by using the point at the center of the stationary Sun!</u> Why did we accept it?

Kepler's first law: (now reductio ad-absurdum)

"The orbit of every planet is an ellipse which has the Sun as one of its foci"
He spoke of the orbital plane of the Sun. (**In Kepler's time the Sun stood still**).
He never knew that the Sun did a spiral orbit, or described a circle.

Kepler's second law:

"The radius vector of each planet passes over equal areas in equal intervals in time.
Radius vector was Kepler's creation.

Kepler's third law:

The squares of the periods of any two planets are proportional to the cubes of their mean distances to the Sun.

Kepler's laws did not explain the behavior of the planets or why he says they move in elliptical orbits, or why their speeds change as he says they do. He never once mentioned in his writings that the Sun moved and formed an orbital track in our galaxy. He never deduced that the Sun moved.

Note: In the universe <u>speed is constant.</u> There is *no slowing down and speeding up* as Kepler stated.

Newton in his book "**Mathematical Principles of Physics**" came up with some answers. He attributed the behavior of the planets to a most fundamental law. **The law of Gravitation.** <u>He too was wrong, wrong because he too, made</u>

the same mistake to use the center of the Sun for his projection, instead of the point at the center of the Sun's orbital Plane. He never knew that the Sun moved, or that it was moving or how it moved. He too was brilliant as he was thought to be, also placed the Sun in a stationary position or on a straight course.

The Axis of the Earth shall point directly to the North Star Polaris constantly while the Sun is at its present location leaving no chance of nutation or wobble of the Earth's axis as Earth moves from point "A" to point "B" between January and July each year. The axial poles must not be mistaken for the magnetic poles, which move slowly away from the axial poles—fading completely every 12,000 years as the Sun makes it round.

Scientist today still believe that the Earth does a sudden summersault. This explanation is an attempt to account for this reversal of the magnetic-poles. Magnetic orientation of the Earth is controlled by the movement of the Sun on its orbit. This is because the axial poles of the planets point in fixed directions relative to other stars in the galaxy while the Sun changes its location. This in turn causes causes a slow change in the distribution of sunlight on the Earth and the planets as the Sun makes its round. The result is that the axial poles are at some times tropical and at other times they are covered with ice. Hence coral reefs in the Arctic and Antarctic belts.

To illustrate how my law works:-

The observer stands at a point **outside our Milky Way galaxy**, he observes the Sun doing a **spiral circle on the inner hub of the galaxy**. He also observes **all planets orbiting the Sun on the same plane in a concentric way—crossing the plane of the Sun's orbit at right angles, regardless of the position of the Sun on its orbit.** He extends their orbits until they intersect or pierce the imaginary point at the center of the Sun's orbit as they move around the Sun. He also observed that with each orbit the Sun makes that the axes of the planets never changed direction relative to other stars in the galaxy, making for seasonal changes on the planets (because of re-distribution of sunlight to new areas).

In the case of the Earth, he saw the degree of tilt of the Earth's axis slowly and constantly changed only relative to the Sun's orbit and still pointed to Polaris as the Sun moved along its orbit. The Sun changed position but axes of the planets remain in a fixed direction. He saw the Sun describe a circle. The Earth's axis pointed to Polaris or Thurban or Vega <u>causing the apparent degree of tilt of the Earth's axis and a corresponding magnetic pole shift increase to 90 degrees when the Sun is at point 9, or Point 3</u>, and to decrease to zero as the Sun gets to point 6, or point 12 and again the magnetic poles are re-aligned with the axis.

He saw that this affected the Earth's magnetic orientation in such a way that the magnetic poles reversed themselves completely with every round of the Sun on its orbit. Hence the theory of summersaulting.
(See drawing figure 3 reversals of magnetic poles).

Some scientists still believe that the Earth does a sudden summersault. **This is reduction ad absurdum.** <u>There is perfect order in the Universe</u>. Pole shift occurs only because of re-distribution of sunlight due to a shifting Sun. **<u>We should no longer have problems forecasting cyclical climatic changes, ice-ages, rising oceans, population explosions, hot-spells and cyclical reversals of the magnetic poles.</u>**

The discovery of an orbiting Sun is the key to true planetary motions. Observation of true planetary motions reveals how all climatic changes occur on the planets in our solar system. This projection shows all planetary orbits intersecting at the center of the Sun's Ellipse. <u>It shows climatic changes occurring in a precise and orderly fashion with no nutation or wobbling of the Earth's axis.</u> <u>It shows the Earth experiencing hot-spells once</u> every 12,000 years, at regular intervals in perfect order. It shows the Sun doing a spiral orbit along the inner hub of our galaxy, once every 25,000 years precisely. Our galaxy is seen three billion light years away from other galaxies.

The oceans of the Earth will be seen rising <u>one **foot every century between ice-ages**</u> that occur simultaneously in both hemispheres whenever the Sun is at point 6 or point 12 on its orbital ellipse on the inner hub of the galaxy.

Chapter 4

THEORIES ON MAGNETIC POLE SHIFT

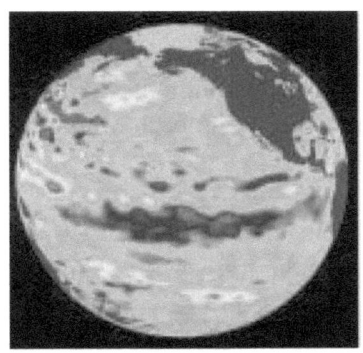

There is no such thing as continental drift, believe it or not. That theory is an excuse for the occurrence of glaciation in the now tropical areas and an attempt to account for the presence of coral reefs now in the arctic. Modern man found it easier to shift the continents on plates in order to account for the presence of these tropical coral reefs that are now located in the Arctic and Antarctic areas of the globe. **Again they are wrong**.

Large areas on both sides of the rift which snakes 47,000 miles around the Earth, do shift because of internal combustion **but that is not continental drift**.

Magnetic Pole Shift

This projection shows the reversal of the magnetic poles of the Earth as the Sun makes its way round on the inner hub of the galaxy:

Note: North axial pole points in fixed direction towards the star Polaris.
Magnetic poles shift as Sun changes position on its orbit.
North magnetic becomes South magnetic as the Sun moves from point 6 to point 12 in 12,000 years.

The magnetic poles are constantly on the move and must not be mistaken for the axial poles, which are in fixed directions relative to the other stars. The observer sees the other planets in our solar system orbiting the Sun on the same plane as the Earth in orderly fashion, all orbits concentric with each other, crossing the plane of the Sun's orbit at right angles at all times—regardless of the location of the Sun. (ad-infinitum).

Movement of the Sun on its orbit determines the movements of the magnetic poles of the Earth. Controlled by sunlight, the magnetic poles shift slowly, reversing themselves completely with each orbit of the Sun. No summersault here, it is a slow process!

In 1982, Peter Warlow, in his book THE REVERSING EARTH, proposes that an extra-terrestrial source is the trigger mechanism for pole shift. He had no idea as to how and why the pole shift occurs. He, like the others, thought that the Earth tumbles at some point in time. Dr. Flodmark in his paper (1981) (THE EARTH'S ROTATION) predicts that the Earth is nearing the moment when another pole shift will occur within the space of a single day. (Preposterous!) They *all think in terms of a summersault or tumbling motion that occurs suddenly*. **They have seen the evidence of these reversals but cannot figure out how they occur.**

In the year 1831, James Clarke claimed that he located the North magnetic pole as far South as the Boothia peninsular. In 1993, the North magnetic pole is located on Bathurst Island moving towards the Northern axial pole. The axial pole is permanently located on the Lamonosov Ridge. This is evidence of a shifting magnetic pole in a given direction towards the axial North pole where they will be aligned with each other when the Sun arrives at point 6 on its ellipse.

Navigators have had to make corrections on their navigational charts to keep up with shifting magnetic poles of the Earth through the passing years. It is being done to this day!

Since the Earth cannot escape the pull of the Sun and go off in a tangent, then it must remain captive and go around it, as the Sun curves in space. As the Sun moves from its position at 6 on its orbit to position 12, the magnetic poles reverse themselves. At points 3 and points 9, the magnetic poles fade. They re-align with axial poles when the Sun gets back to point 6 or point 12. ***North becomes South***.

During this turn-around of the magnetic poles which shift with the moving Sun, the axial poles remain fixed relative to the North Star and other stars in the galaxy. This means that the axial poles must not be mistaken or confused with the magnetic poles. **The timing of these reversals is now predictable. They occur once every 12,500 years. It certainly does.** Studies of magnetic rocks laid at different times in the history of the Earth show that sometimes the North and South magnetic poles are in opposite hemispheres from their present locations. He said "the pattern of repeating magnetic changes through geological time provides a unique calendar against which other events can be calibrated" **They came so close, but never found out!**

If only these eminent scientists had deduced a way, or had they given birth to the idea of ***making all projections from a point at the center of the Sun's orbit while the Sun makes it round on the inner hub of our galaxy***, **then long-term forecasting of ice-ages and catastrophic climatic changes would have been possible earlier in this century!**

John Gribbin also said: "Today it is clear from many similar investigations and other geological evidence that glaciations and ice-ages occur simultaneously in both hemispheres of the globe". (He was right on that). My theory proves that glaciations and ice-ages do occur simultaneously in both hemispheres when the Sun is at point 6 and again at point 12. We have (2) two ice-ages with each orbit of the Sun on the hub of our Galaxy.

Although we have the evidence of pole shift, scientists today still speculate that these reversals take place suddenly as Earth does a "Summersault" (that is utter rubbish).

John Gribbon in his book 'Future Weather' stated and I quote: "Nobody knows for sure how or why the Earth's magnetic field reverses its polarity from time to time but reverse they do". Again he came *so close!*

The Sahara has moved 50 miles South in the last 50 years. This movement is visible along the Sahel Belt of Africa. The North magnetic pole has shifted at a corresponding rate between 1830 and 1990 from the Boothia Peninsula to its present location on Bathurst Island and it continues to move around the Earth as the Sun curves in space. In this way, gradually North becomes South and there is a complete reversal of magnetic orientation on the Earth.

This reversal of the poles takes place over and over in geological time, precisely every 12,500 years. Walter Sullivan said he found evidence that there was a hot-spell on Earth just 6 to 7,000 years ago. He did not know why this hot-spell occurred, nor did he know that it was cyclical. I agree with him entirely as to the timing of the last hot spell which was in fact 6,000 years ago. We know this because in this projection, it takes the 6 to 7,000 years to move between point 9 and point 6. Today the Sun is located at the 6:30 point, where the degree of tilt of the Earth's axis is 23and ½ degrees off the plane of the Sun's orbit.

It also means that at point 9 the axis of the Earth was off the plane of the Sun's orbit by 90 degrees and the Earth's orbit was in line with its own axis, resulting in 6 months darkness and 6 months daylight each year. A year consisted of one day and one night (182 and ½ of regular days when at point 12 and 6.)

Walter Sullivan did not know this. <u>He thought the Sun stood still and made his projections from the point at the center of the Sun</u>. Modern day scientists still come up with the theory of sudden change of polarity of magnetic poles of the Earth suggesting that the Earth tumbles on its axis suddenly. Some scientists who are advocates of tectonics, claim that a spreading sea floor resulting from moving continental plates is the main factor in magnetic reversals of the poles throughout the ages. **This is wrong. The Sun moves in circles. The continents cannot move.**

Evidence of this reversal was found in fossilized rocks and lava several places on Earth and especially in Southern Africa and Australia. Leonard Engel, in recent publication, (**'The Sea'~ Time Life 1973**) stated that the Earth's magnetic field is known to reverse itself every few thousand years so that magnetic North and South poles *SUDDENLY* switch places. He too, was wrong.

<u>They just couldn't figure how it occurred although the evidence of reversal is there.</u>

Some ancient rocks show this reversal in their projection. This happens as the Sun makes its orbit along the inner hub of our galaxy. **It is a slow and ORDERLY process.**

Today, the Earth's axis constantly points to Polaris, at other times to Thurban or Vega, as the Sun makes it round. At the same time the Earth continues to cross the plane of the Sun's orbit at right angles at all times. In this way, sunlight shifts around the Earth and the position of the Sun regulates magnetic orientation on our Earth while the Earth's axial poles remain pointing in a fixed direction relative to the stars.

<u>There is perfect order in this reversal of the magnetic poles of the Earth, once every 12,500 years</u>. The movement is cyclical as the Sun moves. Mr. Engel also stated in his publication that about 450,000,000 years ago, the South magnetic pole occupied the Sahara Desert. <u>He was way off the mark. This happens every 25,000 years.</u> Mr. Edward Bullard a professor of geo-physics said "At regular intervals on the average of every few hundred thousand years, *it flips over* and points South instead of North and down as it does at present in the Northern hemispheres." He too, was wrong when he said : *"it flips over"*.

Chapter 5

PROJECTIONS OF CATASTROPHIC CYCLICAL CLIMATIC MANIFESTATIONS

This new projection also explains the extinction of the dinosaurs, as a result of the long hot and dark spells that overcame the Earth when the Sun gets to point 3 and point 9 on its orbital plane once every 12,500 years. Their extinction is clearly linked to the orbiting Sun. This is truly the dawning of a new day for man. **<u>We made a break-through in forecasting long-term climatic changes on our Earth when we discovered TRUE PLANETARY MOTIONS.</u>**

As a result, we can now plan ahead and get ready for catastrophic changes, like population explosions, hot-spells and ice-ages. During the hot-spells, the Sun is either at point 9 or point 3. While there, the Earth's axis lies in the plane of its own orbit. This causes the Northern and Southern hemispheres to be in complete darkness for 6 months each year alternately. Now that we know this, it gives us time to plan ahead to build our cities on higher ground as the oceans rise by one foot each century between ice-ages.

(London and Venice have gone down one foot per century for the last nine hundred years.) At last no more guessing where we are in the solar system, where we are going and how long it takes to get there and back.

ICE-AGES

Ice-ages occur twice with each orbit of the Sun. That means, an ice-age is on when the Sun is at point 6 and again when the Sun is at point 12. The ice-age is on in both Northern and Southern hemispheres simultaneously when the Sun is at any one of these points.

To illustrate: when traced in colour, the Earth's orbital pattern is seen like a spiral or coiled spring wrapped around the line that forms the orbital ellipse of the Sun.

(see figure 2 illustration in diagrams of coiled spring as the Sun orbits on its ellipse)

This pattern evolves because the Sun is constantly moving along its ellipse which is a very large circle. The Sun's ellipse is so large that if ones stands on it and can see only that small portion, it would appear to be as he is standing on a straight line. However, let us trace the orbit of the Sun in colour, next we number points on the orbit of the Sun as on a clock~ from 1 to 12. Next place a dot at the center of the orbit of the Sun. All planetary orbits on a single plane, must pierce that central point when extended in a concentric way. The orbit of the Sun is so large that it takes the Sun 25,000 years to make one orbit, traveling at 12 miles per second and always at right angles to orbits of other planets in the solar system.

(see illustration of the orbit of the Sun on the clock as it makes its round in 25,000 year cycles, also showing the planets crossing at right angles at all times in concentric way)

(See Illustrations at end of chapters)

To do this projection, take the following steps:

1. Draw a very large circle, I mean as large as possible, to represent the orbital plane of the Sun.
2. Number the orbit points from 1 to 12 as on a clock
3. Show the point as center of the orbit
4. Draw lines from the center of the orbit from all points numbered on it
5. Extend those lines outward from the center to infinity into the celestial sphere.
6. Next we place the Sun at point 6 on its orbit on line with its own equator

The axis is in line with the Sun's orbital plane to Polaris located 57 light years away to the right (a few street blocks away).

* The Earth's axis when extended must pierce the pole star POLARIS.

But its orbit when extended MUST pierce the center of the Sun's orbital plane.

Still at point 6, as the Earth moves from point A to point B in 365 and ½ days, no seasonal changes occur on Earth. It's in an ice-age. Sunlight is distributed equally in both Northern and Southern hemispheres all year round.

At point 6, the degree of tilt of Earth's axis relative to the Sun's orbit is zero. The ice-belts on Earth reach down to 10 degrees North and South of its equator for a period of 3,000 to 5,000 years.

Next, the Sun moves away and goes to a point at 5:30. At that point, Earth's axis still points to Polaris as it is today . . . And 23 and ½ degrees off the Sun's orbital plane. Next we position the Sun at 6:30. The Earth's orbit still crosses the Sun's orbit at right angles. The Earth's axis is again at a point 23 and ½ degrees off the plane of the Sun's orbit, and still pointing toward Polaris the pole star, as the Earth moves from point A to point B, and back to point A between January and July each year.

Prevailing seasonal changes occur on Earth only when Earth's axis attains a 23 and ½ degree tilt, that's when the Sun is at points 6:30; 5:30; 11:30 & 12:30 on its orbit.

Shift of sunlight on the Earth of 23 and ½ degrees North and South of the equator between January and July, causes seasonal changes on Earth as it did when the Sun was at 5:30 o'clock.

NOTE: that as the Earth moves from point A to point B and back each year, that there is no nutation or wobble of the Earth's axis. The axis continues to point to Polaris the North Star directly ~ yet we experience a shift of sunlight 23 and ½ degrees North and South of the equator. A change in the Sun's focus is interpreted as a tilt of Earth's axis. This is not so. The axis is in a fixed position relative to Polaris.

The key to this seasonal change is one of the laws of True Planetary Motions~ that the orbital plane of the Earth shall cross the plane of the Sun's orbit at right angles at all times. As long as this principle is operational, the angle of the Earth's axis relative to the Sun's orbit will continue to increase up to 90 degrees as the Sun moves to a point at 3 on its orbit, and again when the Sun gets around to point 9. At those points, the axis will lie in the plane of its own orbit. This causes the axial poles of the Earth to face the Sun alternately for six months each year. These are the long hot and dark spells that brought about the demise of the dinosaurs.

This period lasts from two to three thousand years. Magnetic pole-shift is then 90 degrees away from the AXIAL pole. To make it clear~ the Sun touches each of the twelve points on its orbital ellipse, while the Earth and the other known planets orbit the Sun on the same plane. All the planets cross the plane of the Sun's orbit at right angles as they follow the Sun on its rounds on the inner hub of our Milky Way galaxy. In this way, no planet can ever over-take the Sun.

The axis of our Earth points to the North Star Polaris as long as the Sun is at its present location.

The axes of all other planets point in fixed directions relative to other stars in our galaxy while the Sun is at its present location. The direction of their axes is independent. The velocity of their rotation and the radius of their orbits are not related to the behavior of our Earth, but they all ride along on the same plane ad-infinitum.

The temperature on the planets is determined only by their distance from the Sun. As the planets cool, Earth will soon be as Mars is today, and Venus will soon be as Earth is today. Mars was exactly as the Earth is today only 5 million years ago. There is still some life on Mars which will be discovered in the next few years. **The pattern of climatic changes on our Earth is orderly and reciprocating. There is perfect order in the planetary motions. Nothing is left to chance or nutation or gyrating or wandering as suggested by Kepler.**

The orbital planes of the planets must pierce or intersect the point at the center of the Sun's orbit at all times.(as seen when traced in colour in illustrations). The conditions that prevail on Earth today return four times with each orbit of our Sun on the hub of the galaxy. Those points are~ 530; 12:30; 11:30 & 6:30. When the Sun gets to any of those points the Earth's axis while still pointing to Polaris is then inclined at an angle off the Sun's orbit by 23 and ½ degrees as the Earth goes around the Sun each year, like it is today in 1992.

<u>Ice-ages are predictable, occurring once every 12,500 years. Ice-ages occur only at two points, point 6 or point 12.</u> At those points, the axis of the Earth lies in the plane of the Sun's orbit with its axis still pointing to Polaris. At those points, the ice belts of the Earth move down to 10 degrees North and South of the equator. The ice-age is on it both hemispheres at those points and lasts for up to three thousand years.

There are no seasonal changes on Earth when the Sun is at those points. The Sun is moving towards the 6 o'clock point today. That means that the next ice-age will be upon us in the next 1,000 to 1,500 years. Today the Sun is located at 6:30 on its orbit moving anti-clockwise toward point 6 on its orbit to the next ice-age! Earth's orbit is still at right angles to the Sun's orbit by 23 and ½ degrees as it points to Polaris. Seasonal changes occur on Earth as it moves around the Sun in 365 and ½ days.

During the last hot-spell, when the Sun was at point 9, on its orbit, the angle of the Earth's axis relative to the Sun's orbit was 90 degrees, with a corresponding magnetic pole shift of 90 degrees, but as the Sun moves on to point 6, the angle of the Earth's axis decreases to zero. At that point, magnetic poles are again aligned with the axis. When the Sun is located at point 6, the Earth's axis and that part of the Sun's orbital plane will lie parallel. The degree of tilt is then zero. Consequently the Earth's orbit will be in line with its own equator, producing equal days and equal nights all year round. **The only survivors on Earth at those times are the ones who move to within 10 degrees to 15 degrees North and South of the Equator.**

The Sun will be at point 6 in 1,000 years. The Earth will be in another ice-age. It is now predictable. Cyclical ice-ages occur with precise timing every 12,500 years as the Sun gets to a point at 6 and 12 on its orbit. Ice closes in on both hemispheres simultaneously.

Between ice-ages, the oceans rise one foot each century. Water from melting ice-caps separate the land masses. Slowly the land bridges are taken by the sea, making it impossible for the dinosaurs, mammoths and other large species to move on, in order to find food as the combination of ice and darkness moved in on them. In time they became extinct. The last land bridges caved in a few ice-ages ago (the very last) after the last ice-age 11,000 years ago.

The rising oceans, the hot-spells and dark-spells, are the direct causes of the extinction of the larger species that roamed the Earth for three billion years. Some scientists believe that the Earth was hit by a meteorite that raises so much dust that the whole Earth was in complete darkness for a long period. There was darkness but from other reasons. The hot and dark spells return whenever the Sun gets to point 3 or point 9 on its orbit.

METEORITES have nothing to do with it!

Man suffered the same fate and came to the edge of extinction several times with each orbit of the Sun. At the peak of the last ice-age, man made his last stand. In Central America and Egypt man managed to survive the last ice-age, but he didn't make it through the hot-spells. He left lasting monuments in those areas. Some of them are still a shrouded mystery.

Modern man is still puzzled as to why they just vanished leaving all their stuff behind them. **Those who did not move on, perished with the changing climate. For example, the nations of the Inca and the Mayans**. There is evidence at the tip of South America that a great civilization flourished there in the past. Their monuments still stand and remain a mystery to modern man. Today only the condors, sea lions and fish can survive there for a long period.

Chapter 6

CYCLICAL POPULATION EXPLOSIONS, RISING OCEANS, HOT-SPELLS AND EXTINCTIONS IN BETWEEN ICE-AGES

Population Explosions

Throughout pre-history and with each orbit of the Sun between ice-ages, there has always been a rise and fall in the population of Earth's inhabitants. We were moving into another billionth human beings on Earth in 1987. These explosions occur before and after every ice-age.

Survival of the larger species:

The larger animals were able to survive for three billion years when there was one continent, "PANGEA". At that time there wasn't enough water on the Earth's crust to form separate land masses. **As the ice-ages came and went, the oceans rose one foot per century. Gradually, the land bridges were gone. The continents were separated. It was not by tectonics or continental drift. The truth is continents never drift. Neither does the Earth's crust telescope into itself. The moving Sun does it all.**

Coral Reefs in the Arctic:

As the Sun curves in space, the Arctic and Antarctic areas become tropical for long periods, and coral reefs developed in places like Spitsbergen in the Arctic circle and in parts of Antarctica which also become tropical~ cyclically.

Hot Spells

There is considerable evidence in pre-history of the hot spells that occur on Earth whenever the Sun gets to point 9 or point 3 on its orbit. Very few human beings survive these hot spells. That is because the axis of the Earth at those times lies in the plane of its own orbit and the sun rises every six months alternately in the Northern and Southern hemispheres.

Extinction Theory

The extinction of the dinosaurs occurred about 60 million years ago when the rising waters isolated them in areas that became totally dark for six months alternately in the Northern and Southern hemispheres.

The extinction of the dinosaurs occurred about 60 million years ago when the rising waters isolated them in areas that became totally dark for six months each year. These hot-spells occur 12,500 years since the formation of the Earth. Before the waters separated the lands, the animals had the sense to move on to the sunlit areas to find food but the rising oceans between ice-ages destroyed the land bridges. Then with no bridges to cross any more ~ the end came for them.

No seasonal changes occur on Earth during the ice-age. This is because the Earth's orbit IS IN LINE WITH ITS EQUATOR when the Sun is at its point 6 or 12 on its orbit. At those points the Earth's axis is also in line or parallel with the Sun's orbit. There are equal days and equal nights all year round. During the high point of the ice-age, the axial pole and the magnetic poles are aligned. Sunlight is evenly distributed in both hemispheres. The ice-belts of the Earth reach down to within 10 degrees of the Equator.

As the Sun moves away from point 6 or point 12, the ice-belts move back to the Arctic and Antarctic. **During the ice-age, the whole of Europe, Canada, the United States, parts of India and Africa, are covered with ice. As in the South, Australia and Antarctica are covered. We found evidence of Oak Forests in the Amazon and perfectly preserved trees in the treeless Arctic and Antarctica. All because of an orbiting Sun.**

Rising Oceans

Between ice-ages, the oceans rise at the rate of one foot per century and continue to rise ad-infinitum.

In the early stages of the cooling process of the Earth's crust, ice formed. Slowly as the Sun made its rounds, the melting ice gave birth to the oceans. Streams formed and became rivers, rivers flowed into lakes which became seas in pre-Cambrian times. Many civilizations were taken in by rising waters between ice-ages. Those who did not move on, just vanished beneath the waters. Atlantis vanished slowly under the rising waters just as Venice is being covered slowly today. Venice has gone down one foot per century in the last 900 years since its construction. All its streets are now under water.

Jacques Cousteau had been exploring and investigating what appears to be man-made walls still standing in some areas of the Earth well below sea level today. He found unexplained submerged columns off the Bermuda Triangle. He also found an area he described as a sunken area off Crete and another off the Azores.

The Bermuda Triangle

We know that the holes off the Bermuda Triangle were made in time by rising waters. Water forced through rocks with the help of daily tidal movements, caused those unexplained holes which Cousteau traced. He poured dye into the currents, got into his driving gear and traced the coloured waters and discovered that the holes in the Bermuda Triangle are all connected by sub-terrainean caves which are formed by tidal water pressure. Deep down in those holes, he found what appears to be fallen columns which also appear to be man-made. This is also an indication that the oceans are indeed rising or there is up and down movements of the Earth's crust.

Man's recorded history only goes back 5,000 years, the reason for this is that 6,000 years ago we came out of the last hot-spell as the Sun moved away from 9 o'clock point. **We moved away from the last ice-age 11,000 years ago when the Sun moved away from point 12**. During the hot-spell, very few living things survive on this Earth. The Sun is at 6:30 point on its orbit and moving anti-clockwise on to point 6 on the orbit into the next ice-age.

Between the hot-spells and the ice-ages, the population explosions occur. Today the Earth's axis is exactly 23.5 off the plane of the Sun's orbit. **The population explosion is on as we enter the next ice-age.** The Sun arrives at the 6 o'clock point in another 1,500 years and the ice-age would be at its height. The ice belts will then be within 10 degrees of the equator. **That will be the time of total disaster. Nearly all living things outside the equatorial belts will perish.** Human beings will die by the millions each day. The only survivors will be those who live within 10 degrees of the Equator, those who manage to be near the axial poles and must live in total darkness 6 months each year until the Sun approaches point 6 or point 12 on its orbit. The opposite happens when the Sun gets to point 9 or 3 on its orbit.

Seasonal changes as occur today on Earth return when the Sun is at 5:30 6:30; 11:30; and 12:30 points on its ellipse. Again at those points the Earth's axis is at an angle of 23.5 degrees off the plain of the Sun's ellipse, resulting in seasonal changes on Earth. As the Earth goes around the Sun at any of those points between January and July and back to January—sunlight shifts 23and ½ degrees North and South of the Earth's equator, causing Equinox, Summer and Winter Solstice.

Always remember that as the Sun makes its rounds on the inner hub of the galaxy, that the Earth's axis always points in the general direction of Polaris, Vega or Thurban. At the same time, the magnetic poles slowly shift to compensate for the movement of the Sun. In this way the magnetic poles reverse themselves completely with each round that the Sun makes, while the axial poles remain fixed.

Chapter 7

THEORIES ON CONTINENTAL DRIFT OR TECTONICS

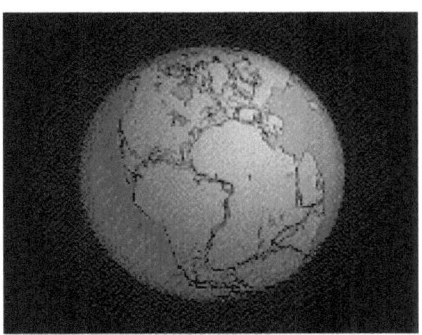

The theory of tectonics or continental drift on continental plates though still widely accepted is in my opinion "Reductio-ad-absurdum". Since the Earth has one solid crust—there could be no such thing as continental drift. No shifting of continents. For example: First let us create a picture of the Earth's crust as one solid shell, with a liquid inner core, "magma". We must drain the water off the Earth, exposing the rifts and ridges and canyons and ancient river beds that lie beneath on the ocean floors and what we see is one solid shell that makes the globe.

The cooling process on Earth was very slow and it continues today. In the early stages of the cooling there was no water, oxygen, parts cooled sufficiently to harden and formed the crust, yet those parts were pushed around with tidal and centrifugal force from internal combustion within the rest of the molten lava or magma. That action caused the foundation of the Himalayas, the Andes, the Rockies, the mid-ocean ridge and rift etc. Internal combustion is still going on today.

The principle is the same as occurs when one bakes a cake. The outer crust hardens and rises from pressure on the inside. Ridges and rifts appear while the inside of the cake is still molten and very hot. In the case of the Earth which is not held in a container, internal pressure causes the rifts and ridges that subsequently snake all around the Earth for a total of 47,000 miles. **The spreading ridges and rifts still occur today and that movement when**

<u>discernable is mistaken for continental drift.</u> **Hence making that theory of continental drift reduction ad absurdum. Cracking and bulging of the Earth's crust still goes on today.**

To this day, there is some up and down movement around Hawaii and Iceland. The pressure from the inside causes the boiling magma to ooze out through the rifts. **There were great upheavals and volcanic eruptions through long periods in time but since the Earth cooled off and in time the entire crust became solidified and ice-caps formed at the axial poles.** The melting ice-caps filled the low areas with water as time went by. There is evidence that some of the areas of the Earth, now above sea level were at times invaded by the sea.

The areas which rose from internal combustion caused the waters to run off as other areas fell in. The sea rushed in and so the low areas were filled with water in time. This accounts for fossilized remnants of sea life found in several places inland on Earth's crust—even in the Andes and the Himalayas. **Occurrences like these gave birth to theories of continental drift.**

<u>The continents never drifted apart and do not drift.</u> As the oceans rise one foot per century, it causes the low areas to fill up with water from the melting ice caps between ice-ages and hot-spells. Whenever the Sun gets to point 6 or at point 12—the ice caps of the Earth measure up to 8,000 miles across and contain 75% to 80% of the Earth's fresh water. This means that with each orbit of the Sun on its ellipse the oceans rise by 240 feet (one foot each century).

CONTINENTAL DRIFT IS A MYTH

It is amazing that in order to account for the coral reefs which developed in places like Spitsbergen in the arctic regions of the Antarctic, they found it convenient to shift the continents instead of the Sun. **So they did just that!**

They moved the continents on plates. In similar maneuver, Kepler invented radius vector, gyration, nutation, aphelion, perihelion and foci etc. when he wrote his laws on planetary motions.

<u>TRUE PLANETARY MOTIONS</u> eluded all of them because they never discovered that the Sun went along in spiral circles on the inner hub of the galaxy.

(With the invention of the "chips", **True Planetary Motions** will eventually be discovered if my projection is rejected.)

Pole shifts brings with it desert encroachment. We are observing desert encroachment as the Sun approaches the 6 o'clock point on its orbit. **In the Sahel belt of Africa man must move South or die.** The desert is moving South at a steady rate of one mile each year. It has moved fifty miles South in as many years, the magnetic North pole is moving toward the North axial pole at a corresponding rate. Desert encroachment is wrecking havoc in Ethiopia. The camps are overflowing—people are dying by the hundreds of thousands.

Mr. Kissinger Secretary of State in the U.S. appealed to the United Nations of the World to help in pushing back the Sahara. There is no turning back. In 1973 200,000 people perished in Ethiopia alone. They have been dying at that rate each year since.

Population explosions occur simultaneously with the carnage that comes with desert encroachment and creeping ice belts, but when the Sun gets to the 6 o'clock point and the ice belts each down to 10 degrees off the equator, **only those people in that narrow strip around the globe will survive**. The population of the Earth now is over five billion since 1987.

The droughts come and go. **The hot-spells come between ice-ages**. After these high points, **as we are at today**—conditions again become intolerable as the Sun gets to point 3 and point 9. Continuous daylight for six months followed by 6 months of darkness.

The heat and the darkness overwhelm the inhabitants of the Earth. **Whole civilizations vanish or perish when the Sun reaches those points. "The Way the Dinosaurs Went".**

The Neanderthals, as a species of early man, disappeared completely as a result of the movement of the Sun. Modern man evolved from the remnants of the surviving Neanderthals between ice-ages. **The population explosion is on. It will be at its height in just 1,000 years.** The demise then begins with a sudden turn. There will be death and hunger all around us. Few of us will survive this period as we move into the ice-age in 1,000 to 1,500 years. **Only those who move to the equatorial belt will survive. This is a WARNING. This FORECASTING is ACCURATE.**

Chapter 8

ICE-AGES, HOT-SPELLS AND THE GREEN HOUSE EFFECT

ICE AGES ARE CYCLICAL

Drilling samples taken from under permanently frozen areas of the Earth, reveal evidence of recurring ice-ages over and over in geological time, **yet most scientists still believe that there were only four ice-ages in the history of the Earth.** There is evidence that the Northern ice-cap passed through the United States and Mexico periodically. The ice still sculptures the rocks and canyons in those areas as the Sun makes each orbit.

Frozen mammoths, remains of deer, oxen, musk, rhinoceroses and tiger-like cats were found in permanently frozen areas of New Siberian Islands, still in the flesh.

Evidence of oak forests exists in Brazil and other parts of the equatorial belt of the Earth. Coral reefs now lie under permanently frozen areas of Spitzbergen. **These areas were indeed tropical once in every 12,500 years. This is because the Sun moves in circles, while Earth's axis points in a fixed direction to other stars.**

Hot Spells

Between the cold glacial times when the Sun is at point 6 or at point 12, there are warm periods. Those periods occur when the Sun is at point 3 or at point 9. During these warm periods, the warmth-loving animals such as the Saber-tooth tiger, Straight-tusk elephant and the Broad-nosed Rhinos, warmth loving plants such as the Fig and Sweet-bay, lived as far North as France and England. There is proof of this. Such evidence is an indication that sunlight does shift around the globe as the Sun circles the galaxy.

As late as 1973 David Bergamini wrote in (Time-Life): "The Universe", that the Sun swings around the galaxy once every 250 million years, and another gentleman, Mr. Mel Hunter in that same book, shows in a painting—the Sun from birth to extinction in 50 billion years.

Of course both of them were wrong.

Up to the time of this writing (1992) no one has come up with the concept of an orbiting Sun on the inner hub of our galaxy completing each orbit in 25,000 years. There is abundant evidence of ice-ages in cycles.

Jane Brody wrote in the New York Times of January 24th, 1974, that this and other finger-lakes in the area were formed 12,000 years ago when the last glacier melted. This lake she said, has been uncovered by Geologists and Anthropologists at the State University of Genesee. She said that woody material found near the top of the lake bed were carbon-dated at 8,000 years, indicating that the glacia formed lake had existed for at least 4000 years. The Coral Reefs found off Spitzbergen now in the Arctic, were once teeming with tropical life-forms.

There is evidence of frozen mammoths in Siberia and Alaska. Coral reefs also in Alaska and well preserved trees thousands of years old, frozen under the treeless Arctic tundra, with leaves and fruit still on them. There is extensive evidence of glaciations and moving glaciers in India, Southern Africa and other equatorial areas of the Earth, even in Brazil, Australia, and China.

The advocates of continental drift conveniently introduced tectonics to account for pole-shift and other catastrophic climatic changes. We must now observe the moving Sun in order to observe true planetary motions. The observer must be placed outside the galaxy.

Rising Oceans

The continental slopes show age with depth. This is conclusive evidence of rising oceans. No explanation is required here. As the oceans of the Earth rise from perpetual melting of the ice-caps, places like Venice, the Everglades, London, Manhattan, Port of Spain, Bangkok and many other areas of the Earth appear to be sinking at the rate of one foot per century. New ocean floors are created as the oceans rise. **Thousands of species became extinct because of rising waters that destroyed the last land bridges between the continents 40 to 50,000 years ago.**

The Last Land Bridges

Towards the end of the last ice-age, 10,000 years ago, the last land bridge between Alaska and Siberia was taken in by rising waters. This caused the greatest loss of life on our Earth in all of time! All of the old routes were flooded and many of the larger species could no longer move to the sunlit areas where they could find food during the long hot-spells followed by dark-spells.

In the past all the continents formed one land mass. The animals moved with the seasons and the shifting sunlight but since the last land bridges gave way to the rising oceans, they either froze or starved to death or were overcome by heat when the Sun reached a point of point three or point nine. In earlier times, they moved freely and evolved over three billion years. In that period they grew to become the largest creatures that ever walked the Earth. The end came slowly, some adapted to sea life, paid the ultimate price of extinction which occurred soon after the bridges continents were gone.

The Green House Effect

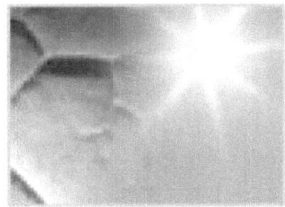

Let us discuss the green house effect and warming trends, as we establish the connection between climatic changes and the orbiting Sun. Global warming on Earth's atmosphere is a serious problem says science advisor at Congress Mr. William Rogers.

The senior N.A.S.A scientist said just two days before Mr. Rogers that the evidence of a warming trend is overwhelming and widespread. (**climatic disruptions are indeed inevitable but for different reasons.)**

N.A.S.A experts said that unless the green house effect is checked, rising temperatures will melt polar ice-caps and this melting will flood low lying coastal areas turning agricultural land into Sun baked deserts, wrecking havoc with rainfall and seasonal patterns all over the Earth. They also blame pollutants for destroying the Earth's ozone layer that screens ultra violet rays. (*Again they are wrong*). **They should be observing a moving Sun.**

Another scientist, Andrew Trevelpiece said that the causes and effects of climatic changes are too poorly understood to warrant changes in energy policy. He is telling us "just put it on HOLD, we don't know enough about this to change and rearrange policy on energy". Those in Japan said they saw no need for international controls. Third World countries showed no interest in checking the green house effect.

I am saying that whether or not they decide to act on it one way or the other, They are all on the wrong track.

The effect that pollutants have on long term or short-term climatic changes is so *infinitismal* that whatever man does cannot and will not cause the slightest ripple in the winds of change relative to climatic changes on Earth. **We must turn our attention to the orbiting Sun. We are looking the wrong way,**

at the green house effect and global warming. For example, the Sun goes around in circles, the Earth's axis point to Polaris 57 light years away, the Earth's orbit crosses the plane of the Sun's orbit at right angles at all times and when Earth's orbit is extended in a concentric way it touches the point at the center of the Sun's ellipse at all times.

Since the industrial revolution we have evidence that temperatures in the U.S and Northern latitudes have actually decreased over the last half century, yet N.A.S.A presented a picture to the senate committee of unqualified gloom and doom by the year 2030.

We have evidence that mean U.S temperatures have actually dropped in the last 30 years!

Scientists are now asking why isn't the Northern hemisphere warming if the amount of Co2 in the atmosphere is increasing? The best explanation is that other factors unknown or currently unexplainable are interfering. There is also evidence that the Southern hemisphere has been warming up in a fashion that no one doubts. They are now asking why the difference between hemispheres. (no problem, the morning Sun is the answer).

The people at N.A.S.A are now saying "until we understand why the northern hemisphere has been so slow to warm, please pass the research funding and don't sell the condo at Ocean City for fear that the rising oceans caused by melting ice caps will cover you. In other words, they spoke at length on a subject of which they know little or nothing.

The theory on the green house effect is not entirely for the trash can, it should be put in a class with the theory of tectonics or continental drift. A few years from now we will be looking back and saying how dumb we were in 1990. It is evident that the United States Senate and Congress and N.A.S.A are deeply concerned with the green house effect on our Earth, which they say is created by pollutants.

Goddard Space Flight Center official James E. Hanson, before a Senate panel said in 1986 that global temperatures will rise to a level which has not existed on Earth for the past 100,000 years because of man's pollutions. **That is pure speculation**. He went on to say that it may not be too late to avert major climatic changes. According to N.A.S.A calculations, by the year 2020 the average temperatures across the U.S will rise by 9 degrees Fahrenheit or more. According to Stephen Leatherman of Maryland University, sea-levels have risen one foot in the last century. **He was right**.

We have evidence that the waters around Venice have risen 9 feet in the last 900 years since its construction. Not because of pollutants but because of the rising oceans caused by melting ice caps, as the Sun makes its round on the inner hub of the galaxy while the Earth's axis continues to point in the direction of Polaris. **Contrary to popular belief, Venice is not sinking, the waters are rising.** Sherwood Rowland a University of California chemist warns that with the green house effect going on indefinitely, we will have temperature rise that will extinguish human life within 5,000 to 10,000. I say rubbish, reduction ad absurdum.

We are moving into an ice-age as the Sun moves from 6:30 to 6:00 point in 1,000 years. In this projection, I am using the point at the center of the Sun's orbit. We can now, from this vantage point, predict ice-ages that occur every 12,500 years whenever the Sun gets to a point at 12 o'clock or point 6 o'clock on its orbital plane.

This happens cyclically all through the history of the solar system. The discovery of true planetary motions has eluded man through all time because a moving Sun was never considered when trying to determine true planetary motions. We are on the brink of another population EXPLOSION as we enter the next ice-age. It has already began.

Chapter 9

ACCURATELY FORECASTING CYCLICAL GLOBAL CLIMATIC CHANGES

TIME SCALE

Here is a time scale which will aid in long term forecasting.

Since the Sun makes one orbit in 25,000 years, it would take ½ that time to move from point 12 to 6, which is 12,500 years. **The positions are numbered as on a clock, from one to twelve. Each hour on this chart, represents 2,083.33 years of travel by the Sun. The sequence is as follows:**

      THE SUN IN ITS 25,000 YEAR ORBIT:
- ➤ 12 o'clock= ice-age
- ➤ 9 o'clock=hot-spell
- ➤ 6 o'clock=ice-age
- ➤ 3 o'clock= hot-spell

Back to12 o'clock to complete the cycle. A total of 25,000 years.

In the mean time the axes of all the planets in the solar system point in fixed directions relative to the other stars in the galaxy. Today we know that the Earth's axis points to Polaris and lies at an angle of 23 ½ degrees off the plane of the Sun's orbit's. Conditions prevailing on Earth today do so only because Earth's axis is off the plane of the Sun's orbit by 23 ½ degrees.

As the Sun makes its round, this situation develops at four points, mainly 11:30, 12:30, 5:30 and 6:30. Since we know that we must come out of a hot-spell about 6,000 years ago, and these hot-spells can only occur at point 9 or point 3, this means that the Sun can only be at one of two points today moving anti-clockwise. Those two points are 6:30 or 12:30. At any of those two points we are certainly moving into an ice-age which will occur in both Northern and Southern hemispheres simultaneously.

There is no recorded history covering the period before the last hot-spell.

Man's recorded history begins just as we came out of the last hot-spell 5,000 years ago around 3,000 B .C. Our records begin just emerging from the last hot-spell when the Egyptians built pyramids. The Sun had just moved away from point 9 on the chart. Let's assume that around 3,000 B.C that the Sun had just moved away from that point, then 2,000 years later, in the year 1,000 B.C., **moving anti-clockwise** the Sun would have been at point 8, and in the year 1,000 A.D, 2000 years later, the Sun was at point 7. (Christ arrived at point 7:30 on the chart.)

The angle of the Earth's axis relative to the Sun's orbit increases and decreases as the Sun curves in space because of the fixed positions of the axes of the planets. At point 9 the degree of tilt is up to 90 degrees. At point 6 or 12, the degree of tilt goes down to zero. In this way the magnetic poles of the Earth reverse themselves completely each time the Sun makes its round. This reversal takes place only because the axis of Earth points in fixed direction relative to other stars in the galaxy as the Sun makes it its round.

It is 1993 A.D. The Sun is at 6:30 and Earth's axis is 23 ½ off the plane of the Sun's orbit. In the year 2993 the Sun will be at point 6, and the ice high point will be at both hemispheres simultaneously. **This scale makes forecasting possible. Today we are already 1,000 years into the present ice-age. The high point will occur in another 1,000 years, and by the year 3,990 conditions on the Earth will again prevail as they are today. By 3,990 the Sun would have moved to the point at 5:30. The Earth's axis will again assume an angle of 23 ½ degrees off the plane of the Sun's orbit.**

There is abundant evidence that we moved away from the last hot-spell 5,000 years ago, or 3,000 B.C. The events follow in sequence. Things get very bad during the hot-spells. It is six months darkness and six months sunlight as Earth's axis gets in line with its own orbit and still crosses the plane of the Sun's orbit at right angles. For 6 months the North Pole faces the Sun and the next 6 months the South Pole faces the Sun.

Very few living things survive these hot-spells, World population drops to a low point each time. The ice-caps melt away completely and oceans rise. New breeds of human beings evolved between ice-ages and hot-spells. Between 3,000 and 1,000 B.C. seasonal changes became discernable once more. The Neanderthals became extinct somewhere along the way. The angle of the Earth's axis relative to the Sun's orbit decreases from 90 degrees at point 9 to about 45 degrees when the Sun gets to point 7:30 and down to zero at point 6.

By the year 1,000 A.D. there was considerable rebirth of life on Earth increasing steadily up to today where human population is already over 5 billion! We are in the middle of a population explosion. The Earth's axis is again at 23 and ½ degrees off the plane of the Sun's orbit still pointing to Polaris because the Sun has moved down to 6:30 moving anti-clockwise.

Let us review the angles of the Earth's axis relative to the Sun's orbit as the Sun makes its rounds:

Hot-spells occur only when the Earth's axis is in line with its own orbit and at right angles to the Sun's orbit.

As the Sun moves to point 9, the Earth's axis is again 90 degrees off the plane of the Sun's orbit and the Earth's orbit is again in line with its own axis which still points to Polaris or Thurban. Same as when the Sun was at point 3.

This makes forecasting simple. The timing is orderly and precise. We know that the Sun makes one round in 25,000 years. That is the time it takes the Sun to move from point 12 to point 6. **Therefore by determining the position of the Sun on its orbital plane which can be pin-pointed by observing the degree of tilt of the Earths' axis relative to the Sun's ellipse, precise forecasting of ice-ages is now a reality!**

The observer discovers that there is no nutation or wobbling of the Earth's axis. No summersaulting of the Earth, resulting in the SUDDEN reversal of Earth's magnetic poles.

The observer sees the reversal of the poles taking place slowly as the Sun moves between point 12 and point 6. He sees no precession of equinoxes, no retrograde motions of the planets.

No continental drift, no radius vector, no aphelion or perihelion.

1. At point 12, the Earth axis lies in the plane of the Sun's orbit and points in the direction of Polaris and Vega. The degree of tilt is then zero.

2. Next, at point 12:30 the Earth's axis is off the plane of the Sun's orbit by 23 and ½ degrees, and points to Polaris and Vega. The degree of tilt is then 23 and ½ degrees.

3. Next, the Sun moves to point 3 and the Earth's axis is off the plane of the Sun's orbit by 90 degrees and still points in the direction of Polaris, Vega or Thurban. Its orbit is in line with its axis. The degree of tilt is 90 degrees, relative to the orbital plane of the Sun.

4. Next, as the Sun moves to point 5:30, again the angle of the Earth's axis relative to the Sun's orbit is off by 23 and ½ degrees and still points to Polaris. The degree of tilt is 23 and ½ degrees.

5. Next, the Sun moves to point 6 and again the Earth's axis lies parallel to the Sun's orbit. Earth's orbit is in line with its own equator and still crossing the Sun's orbit at right angles. The degree of tilt is again zero. Earth's axis is still pointing towards Polaris and the ice-age is again on, at both hemispheres.

6. Next, the Sun moves on to point at 6:30. Here we find that the angle of the Earth's axis relative to the Sun's ellipse is again as they were when the Sun was at 12:30, 5:30, and 11:30.

After each hot-spell as the Sun leaves point 9 or point 3, there is a rebirth on Earth. We began our present era 5,000 years ago after the last hot-spell. Since then the population has risen at an alarming rate.

In the last 30 years, world population has doubled to over 5 billion. A dramatic increase was observed 500 years ago and the trend is continuing even among the animals. Some species will grow smaller in size and millions will starve to death.

Eventually, after a few more orbits of the Sun on the inner hub of the galaxy and a few more ice-ages, some may have to migrate to Venus or adapt to sea life as the Earth cools and takes on a look of Mars today. Our future is within the sea or on to Venus.

Chapter 10

VANISHING CIVILIZATIONS, SURVIVAL AND MOVING INWARDS IN OUR COSMOS

Migrating Inwards

Any migration to another planet must be inwards in our solar system. It is already too late for us to go to Mars. All the outer planets including Mars, are already too cold. Planets cool off relative to their distance from the Sun. Mars has already passed though the cooling period that the Earth now enjoys. Remnants of Earth-like life forms still exist on Mars today. Soon in the next few years life will be discovered on Mars.

Life flourished on Mars two billion years ago, just as it prevails on Earth today. Mars has cooled off considerably because of its distance from the Sun. Most living species on Mars in whatever form, have already become extinct because of intolerable cold and alternating hot and cold dark spells.

Most of the water on Mars is now in ice form. The ice-belts are now within 10 or 15 degrees off the equator and its orbit is in line with its equator. **It is in an ice-age**. They are experiencing equal days and equal nights all year round. The orbit of Mars continues to cross the orbital plane of the Sun at right angles.

Mars is now in an ice-age. All water on Mars is locked up in frozen ice-caps which reach down to its equator.

Back to Rising Oceans on Earth

If one searches the ocean floors of our Earth by drawing off the water, the observer will find the remains of sunken cities that existed in pre-history when the level of the oceans was much lower than it is today.

Atlantis must be there somewhere off Europe in the Atlantic or the Mediteranean or the continental shelf off the Canaries. All areas of the Earth that are less than 20 feet above sea level today, will be under sea level in just 3,000 years. Man walked the floor of the Indian Ocean and all the continental shelves of the Earth before the oceans covered them in time. Aged bones of human beings found directly in Ethiopia and Zambia reveal evidence that man walked the Earth as far back as 3 million years ago.

Dated pottery shows that they were highly civilized even then. Fossil beds discovered in the Americas recently prove that pre-historic man roamed the Americas more than three million years ago and that they were overcome by catastrophic climatic change over and over in geologic time.

The waters in places off the tip of Asia and North America are already 60 feet deep. This means that at the rate at which the oceans are rising, that a land bridge existed between these two land masses very recently in pre-history, about 7 to 10 thousand years ago. The oriental race or yellow race roamed Asia and the Americas for more than two million years. Great civilizations developed and flourished in the Americas for thousands of years.

Throughout Central and South America, monuments constructed 3,000 years B.C. still stand, and some of them only 28 miles from Mexico City. The ruins of civilizations of pre-history can be seen in Peru and Columbia, etc.

We can find the answers to these vanishing civilizations only by observing the orbiting Sun. Man has always made the same mistake by not moving with the orbit of the Sun, the moving Sun causes catastrophic climatic changes on Earth each time it goes around the inner hub of the galaxy. Shifting Sunlight or the re-distribution of Sunlight over the Earth, makes all the difference.

Life on Earth

Life forms first appeared on Earth in pre-Cambrian times about four to five billion years ago. The Earth in its early formation was a globe made up of boiling lava. It cooled slowly, the crust slowly formed. In time, ice formed at the far ends of the axis and with each orbit by the Sun, the ice melted. Waters rose and triggered the beginning of early one-celled life forms that evolved into soft bodied animals etc.

Life in the Cosmos

We are not unique in the cosmos, given enough time and the right temperatures, materials develop all over the cosmos in the same way. The process repeats itself on other planets located in billions of other galaxies and solar systems in the cosmos. As long as a planet is cool enough and the cooling process is slow enough as to appear stabilized long enough, life forms can and will develop in like manner as they developed on our Earth. There is great possibility that life exists elsewhere in our universe on billions of other planets which became trapped on orbit around other stars.

Planets form as stars explode and magma is strewn out in space. Magma became trapped in orbit around other stars forming new planets. In time they too cool off and when they do, conditions prevail on some of them just as they do on Earth today. There will come a time when our Earth will be as cold as Mars is today and at another time as cold as Jupiter and the others farther away from the Sun. There will come a time when Mercury will be the only planet in our solar system capable of sustaining life forms as exist in the Solar System. **It is already too late to go to Mars!**

Chapter 11

TRACING THE CONTINENTS BY CHARTING THE OCEAN FLOORS

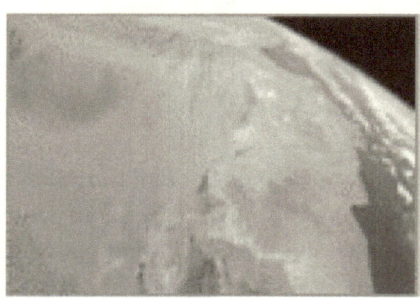

Mapping Ocean Floors

We are now able to map ocean floors and we are discovering more evidence of rising oceans on the floors of continental shelves all over the globe. We can trace river beds for hundreds of miles on the ocean floor. River beds that were once above sea level but surrounded to the rising oceans. For example: the Congo river in Central Africa can be traced out into the Atlantic. It has been traced for hundreds of miles out on the ocean floor.

The Amazon River in South America has been traced out into the Atlantic on the continental shelf.

The Mid Ocean Ridge and canyon were traced south of Greenland into the mid Atlantic. The Mississippi in North America has been traced on the ocean floor of the Gulf of Mexico to the Mississippi Cone.

In the Bay of Bengal, there is a clear follow up of the Ganges River and its tributaries on the ocean floor beyond the Ganges cone. On the other side of India—the Indus River bed runs for over 1,000 miles on the Arabian Sea floor.

With modern equipment, these river beds and canyons, rifts and reefs can now be traced in all areas around the globe that are below sea level.

The deep gorges and canyons found cut into the face of the continental slopes should not puzzle Oceanographers as they do. They want to believe that some great force had carved them. Even today, they suggest that they were cut by powerful under-sea currents or torrents of silt laden waters or turbidity currents. This is not the case. These river-beds and canyons and under water gorges are nature's way of recording its past. These were cut by rushing waters from melting ice-caps and glaciers in pre-history before they too were covered by the rising oceans.

Similar gorges and canyons can be located on Mars today, but the waters no longer flow, because its orbit is in line with its equator at this point in time, this causes the ice belts on Mars to close in on its equator. In other words, Mars is in an ice-age.

Most of the East Siberian Sea, Chuck-Chi Sea and Bering Sea, were above Sea level 25,000 years ago. The Chuck-Chi Range was part of the Rockies but became separated by rising waters. The Mid Ocean Ridge and Mid Ocean Rift both took their shape as the Earth's crust hardened. The turbulence of the molten magma on the inside of the crust, caused by boiling and centrifugal force and inertia as the Earth spun on its axis, produced this Mid Ocean ridge and rift that snakes around the Earth for 47,000 miles.

The rift runs along the center of the ridge and can be compared to the rifts and ridges formed on baked cakes. E.g. the cake is soft on the inside as the crust hardens while it bakes in the oven. There is pressure on the inside as the crust expands and is pushed upwards and outwards.

First, a ridge appears, then the ridge splits to form a rift along the center of the ridge on the cake. This is precisely what happened on our Earth in the early stages of the hardening of the crust and general crust formation as the Earth cooled. In the case of the cake—it is in a container and can only expand at the top but in the case of the Earth, the pressure comes out on all sides. We see the ridges and rifts snaking around the globe for 47,000 miles through every ocean on the Earth's surface.

Sea Floor Ages with Depth

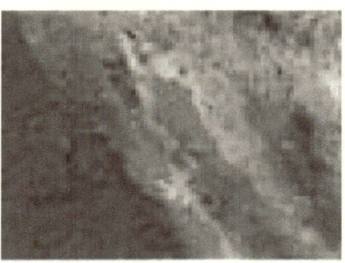

The sea floor ages with depth or with distance from the ridge. It is proof that the oceans are rising.

Let's go back to tectonics or continental plates and continental drift. It is still widely accepted that these plates move apart in sections called plates and that these plates move apart or into each other, tearing continents apart, carrying land masses with them. That is purely speculation, **there is absolutely no evidence of this shifting of continents.**

Thirty years ago the theory of continental drift was dismissed as a whimsy of man. Man had a problem of explaining the appearance of similar fossilized creatures on continents now separated by oceans. He found similarities in geological formations on parts of continents that are now separated. He found fossils of ancient creatures such as the Mesosaurus that could be found only in Brazil and across the Atlantic in South Africa.

Alfred Wegner published a book in 1915, entitled "The Origin of Continents and Oceans". Most American Scientists scorned the Earth-shattering idea that continents could glide across the surface of the Earth. Wegner was indeed wrong. He died in 1930.

The Continental Drift Theory was in disrepute for decades. Continental Drift is a myth (reducto—ad absurdum)

Only an eye-blink ago, the answer was (ABSURD), yet now there is a prompt acceptance of what was once heresy. Man is now introducing continental drift which is a retrograde step. Man moves the continents in order to explain the presence of coral reefs in the arctic off Spitzergen and other areas in Antarctica. He provides an answer by moving the continents on plates.

This is because he totally disregards the movements of an orbiting Sun which bring about all climatic changes on Earth because of the manner in which Sunlight is distributed over the Earth as the Sun travels along its orbital ellipse on the inner hub of our galaxy, once every 25,000 years~ ad Infinitum.

Today scientists like Carl Sagan and others, are just beginning to entertain the idea or concept of an orbiting Sun in our galaxy. Carl Sagan in his book "Cosmos" hinted that our Sun is a star on a spiral arm of our galaxy.

My projection on planetary motions is done from a point at the center of the orbital track, formed by the Sun as it orbits the inner hub of our galaxy. **I am truly amazed that no one on the history of the time of man on this Earth, ever deduced such a theory that shows true planetary motions as the planets orbit the Sun, crossing the plane of the Sun's orbit at right angles at all times while their axes point in fixed direction relative to other stars.**

RISING OCEANS

Coral polyps record rising oceans. The rise of the oceans is recorded by polyps.

As the oceans rise their age can be determined by coral polyps that are known to build on what have been called sinking volcanoes in the Pacific. Some of them had begun building millions of years of years ago at their favourite level of 150 feet. Even as the oceans rise, they continue to build at 150 feet below sea level. They keep moving on up, as oceans rise, always at 150 feet below sea level. Today, after millions of years since they began building, modern man drove shafts through layer after layer of coral at Einwetock Atoll and struck solid volcano of which these polyps began building at 4,220 feet. Since they did build as far down as 4,220 feet but always lived at 150 feet below sea level, it seems that they moved up slowly with the rising oceans. These polyps lived at all times at 150 feet, yet they were found stacked on each other at 1,220 feet below.

Even Charles Darwin identified them as sinking volcanoes. It did not occur to him that the oceans are rising. In the year 1837 Darwin came close to discovering that the oceans are rising, with his brilliant surmise—he suggested that if one dug far enough, one would find the sunken volcanic island on which polyps had began building millions of years prior. It only occurred to him that the islands sunk. He never deduced that oceans could be rising.

Having struck volcanic rock at such depths today—it means that at the time that these polyps began building on this atoll that the ocean level was about 4,000 feet lower than it is today. In Darwin's time (1837)—the North magnetic pole was located on Boothia Peninsula but it is now located on Bathhurst Island (1911). The magnetic poles have shifted as many miles in as many years.

Chapter 12

EXTINCTION OF THE SPECIES AS A RESULT OF THE ORBITING SUN

THE EXTINCTION THEORY

The mystery of vanishing species:

We have already found evidence of man's habitation in pre-history of permanently frozen regions of the Arctic. Anthropologists found flints and other signs of human habitation more than 10,000 years old on the frozen shores of the Arctic in Northern Canada. We know that man could not live on a frozen shore where he could not find food. The fact is that those areas were indeed equatorial areas just 10,000 years ago because of the moving Sun. Man walked across the land bridge where the Bering Sea is today. He roamed the Americas for over two million years.

Man and beast walked the floors of the oceans which lie under water in the Indian Ocean and Pacific Ocean. It is believed that 50,000 species of large mammals vanished from the Americas about 10,000-40,000 years ago. To date it is still a mystery as to what killed them. *We now have the answers.*

Here is the answer:

The end came for them as the rising oceans separated the land masses with each orbit of the Sun. The weight of the large mammals dropped 80-90% in that short space of time. The animals included horses, camels, sloths as large as elephants, mammoths, vast herds of giant bison and rhinos. There were several extinction theories which included sudden climatic change, over kill by primitive man, epidemics etc. Some scientists now wonder, not about how they became extinct, but how they managed to survive for thousands of years. The extinction should no longer be shrouded in mystery. Some of the factors are—rising oceans, ice-ages and hot-spells. **<u>All the result of an orbiting Sun.</u>**

Some of the animals mentioned, existed up to 10,000 years ago in Siberia and Alaska or Chile and Peru. In some cases the deserts pushed them to extinction just as the ice did in the ice-age or the hot-spells with long periods of darkness. The desert is doing it today to man in the Sahel belt of Africa. In the case of the animals, they became more and more isolated as the oceans rose, taking in the land bridges. Some species reduced 80-90% in size in a period of 3,000-4,000 years.

With the land bridges gone, they could no longer roam the Earth as they pleased. They could no longer move away from the ice, the hot-spells or the darkness. They were trapped in dark areas where they could not see to find food. The steep drop in weight of the larger species came in the last 10,000 to 40,000 years. Thousands of species were trapped in Alaska and the tip of South America. Many maybe found fossilized on the ocean floors and on the continental shelves, especially on the North slopes off Siberia.

Man has no control over the cyclical fluctuations and periodic droughts. The Sun moves in an orbit on the inner hub of our galaxy and it is this factor that governs the dimensions of desert encroachment all over the Earth. One sure way for man to accelerate the cooling process on this Earth, is by tapping the inner core from thermal heat. That will be his greatest mistake. It will take some time but that will speed up his demise. Then we'd have to get off on Venus. To survive, we must move inward in the solar system. The outer planets are already too cold.

Chapter 13

THE MISSING PUZZLE PIECE: SCIENTISTS HAVE COME SO CLOSE IN THE PAST BUT TRUE PLANETARY MOTIONS HAS ELUDED THEM ALL

FORECASTING A PROBLEM

In the year 1842 Joseph Adhemar a French mathematician started people thinking about how changes in the Earth's orbital geometry might indeed cause ebb and flow of great Northern hemisphere ice-ages. The idea led to an accepted theory of the ice-ages interglacial rhythms every 11,000 years. Adhemar was right when he said that the Earth was in the grips of an ice-age 11,000 years ago and that the one before was 22,000 years ago and 33,000 years ago throughout time. He knew of this cyclical rhythm, yet he failed to devise the correct projections.

He was right about the timing of the ice-ages but he never knew what caused the cyclical rhythm. This eluded him because he didn't know that the Sun went around in circles on the inner hub of our galaxy. Although Adhemar was right, he was criticized by men like Humboldt and others who didn't make any better contributions on the subject of ice-ages and long term climatic changes on Earth.

After Adhemar, James Croll in the 1860's, started us off on the correct path to an astronomical theory on ice-ages. As a janitor at the Anderson College and museum in Glasgow he shook the scientific world with his brilliant theory of ice-ages. He left school at 13, yet along the way he found time to complete a major book of his own entitled 'The Philosophy of Theism'. He became a well known scientific thinker and from 1867 onwards, James Croll was accepted as a scholar of distinction by the scientific world. He also wrote 'Climate and Time', which was a summation of ten years of his work on the cause of ice-ages. He was elected as a Fellow of the Royal Society in 1876. Croll had a knowledge of the cycle of the precession of Equinoxes and linked it to the tilt of the Earth's axis and orbital eccentricity. His idea was a complete shift from Kepler's theory on planetary motions. **Croll discovered that the Earth's orbit is almost a circle and not the ellipse that Kepler projected**.

Kepler used the Sun as one of its two foci, perihelion, aphelion and radius vector etc. **Of all the modern astronomers from Kepler to Carl Sagan et al, James Croll and Joseph Adhemar came the closest to discovering true planetary motions and long term forecasting of ice-ages and interglacial etc**. Another astronomer, Milankovitch, confused things once more with his theory. He came up with extensive calculations of radiation curves. In 1930 he suggested that it is a reduction of summer isolation and not the occurrence of severe winter that is the key to ice-ages. But by 1950 new evidence seemed to have left the Milankovitch model out of step with the REAL world and was rightly discarded by all but a handful of Climatologists.

MAGNETIC REVERSALS

John Gribbin in his book 'Future Weather' stated and I quote: "Nobody knows for sure how or why the Earth's magnetic field reverses its polarity from time to time, but reverse it certainly does." Studies of magnetic rocks laid down at different times in history of the Earth show that sometimes the North and South magnetic poles are in opposite hemispheres from their present locations." He said also "the pattern of repeating magnetic changes through geological times provides a unique calendar against which other events can be calibrated" end of quote.

If only these eminent scientists had deduced a way or had they given birth to the idea making all projections from a point at the center of the Sun's ellipse while the Sun made its round on the inner hub of our galaxy, the long term forecasting of ice-ages and catastrophic climatic changes would have been possible earlier in this century. John Gribbin also said, I quote: "today it is clear from many similar investigations and other geological evidence that glaciations, ice-ages, occur simultaneously in both hemispheres of the globe." This rules out the Milankovitch theory and supports my theory.

FINAL THOUGHTS

The discovery of **TRUE PLANETARY MOTIONS** has eluded man all through time. The day of long term forecasting and predictions of catastrophic changes is now here.

We can now **prepare for cyclical population explosions, rising oceans, glaciations, ice-ages and hot-spells.** We can now adjust navigational charts to keep up with the shifting magnetic poles of Earth as the Sun makes its round. **The theory of shifting continents and tectonics is now reduction ad absurdum.**

Finally I must stress the factors which govern the changes in the orbital geometry of the Earth that must be considered:

1. The position of the Sun in the inner hub of our galaxy

2. The use of the point at the center of the Sun's orbital plane in making projection relative to planetary motion.

2b. There shall be no wobble or gyration or nutation of Earth's axis or axes of other planets in the galaxy. All axes of the planets **MUST** point in fixed directions relative to other stars as the Sun makes its round. There is not tilting of axes as our Sun moves from point to point on its orbit. **There is perfect order in the universe.**

3. The fact that all planets orbit the Sun on a single plane ad-infinitum while at the same time crossing the plane of the Sun's orbital at right angles at all times with their axes in fixed directions relative to other stars in our galaxy.

This is the key to **TRUE PLANETARY MOTIONS**. This discovery unravels mysteries of shifting magnetic poles, ice-ages, catastrophic climatic changes, cyclical population explosions, extinction of the species, rising oceans, etc.

This is truly the dawn of a new day, a new beginning for man! Our survival into the future depends on heeding the warnings herein now made possible by the projections of this discovery of the century.

CHARTS, DIAGRAMS AND REFERENCES
TO TRUE PLANETARY MOTIONS

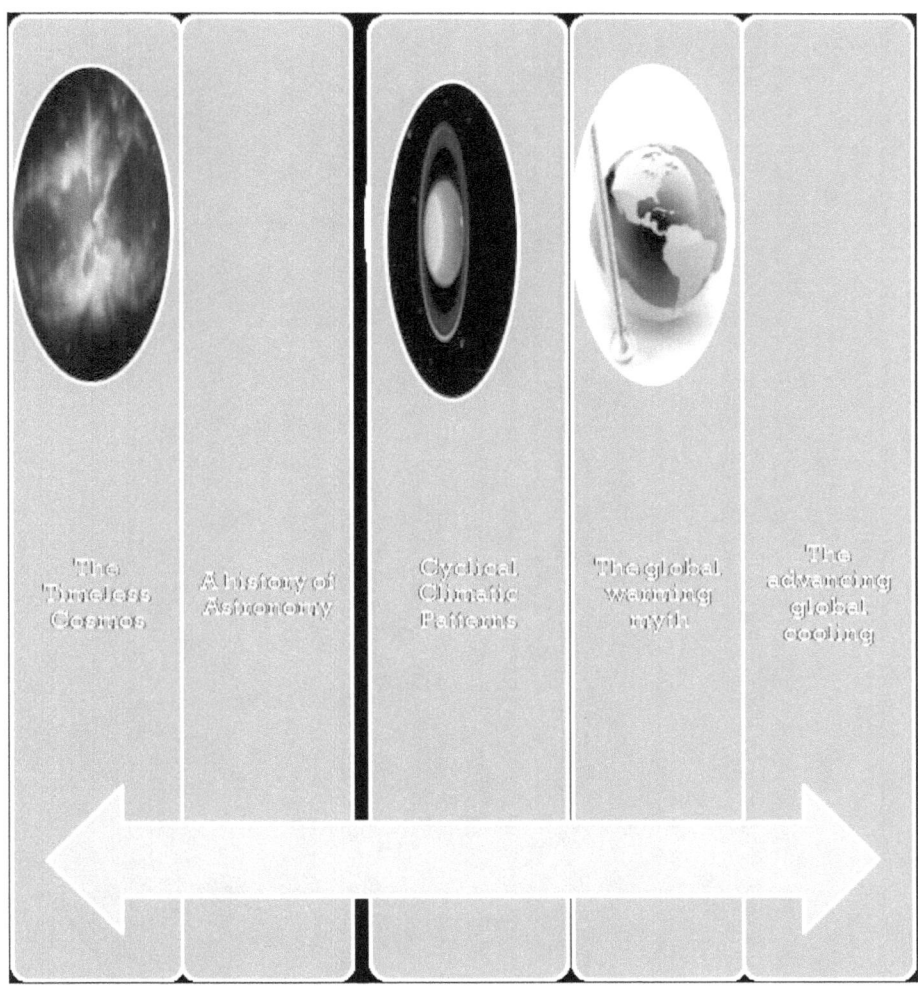

The anatomy of the discovery: Galaxies

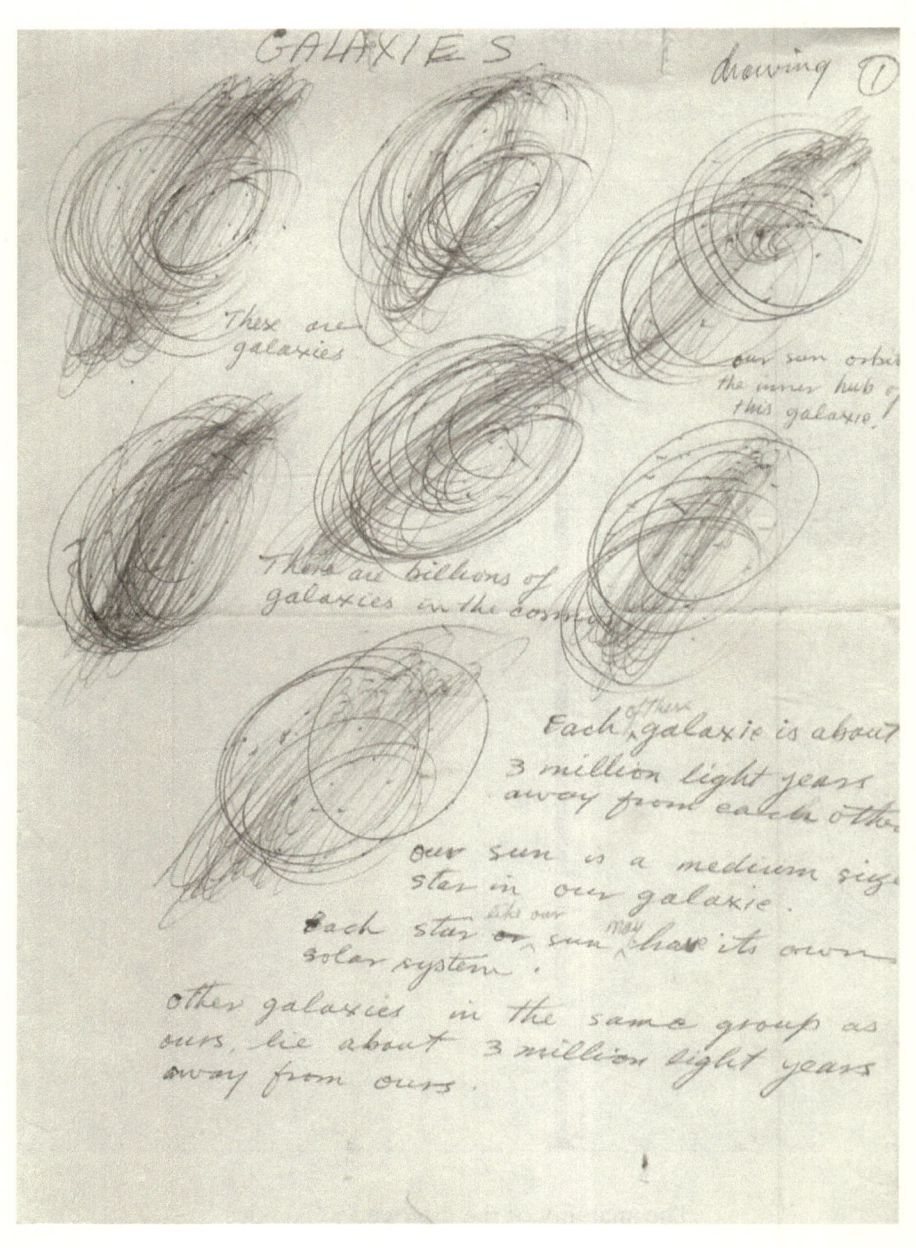

Figure 1: The orbit of the Sun

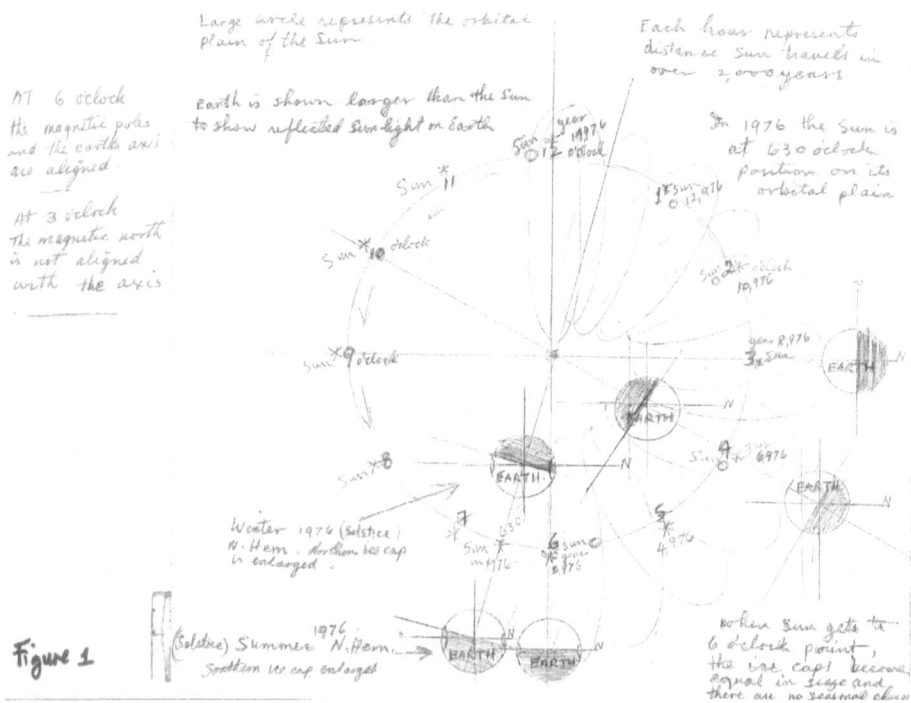

Figure 1

Figure 2 : Observer's Point of View

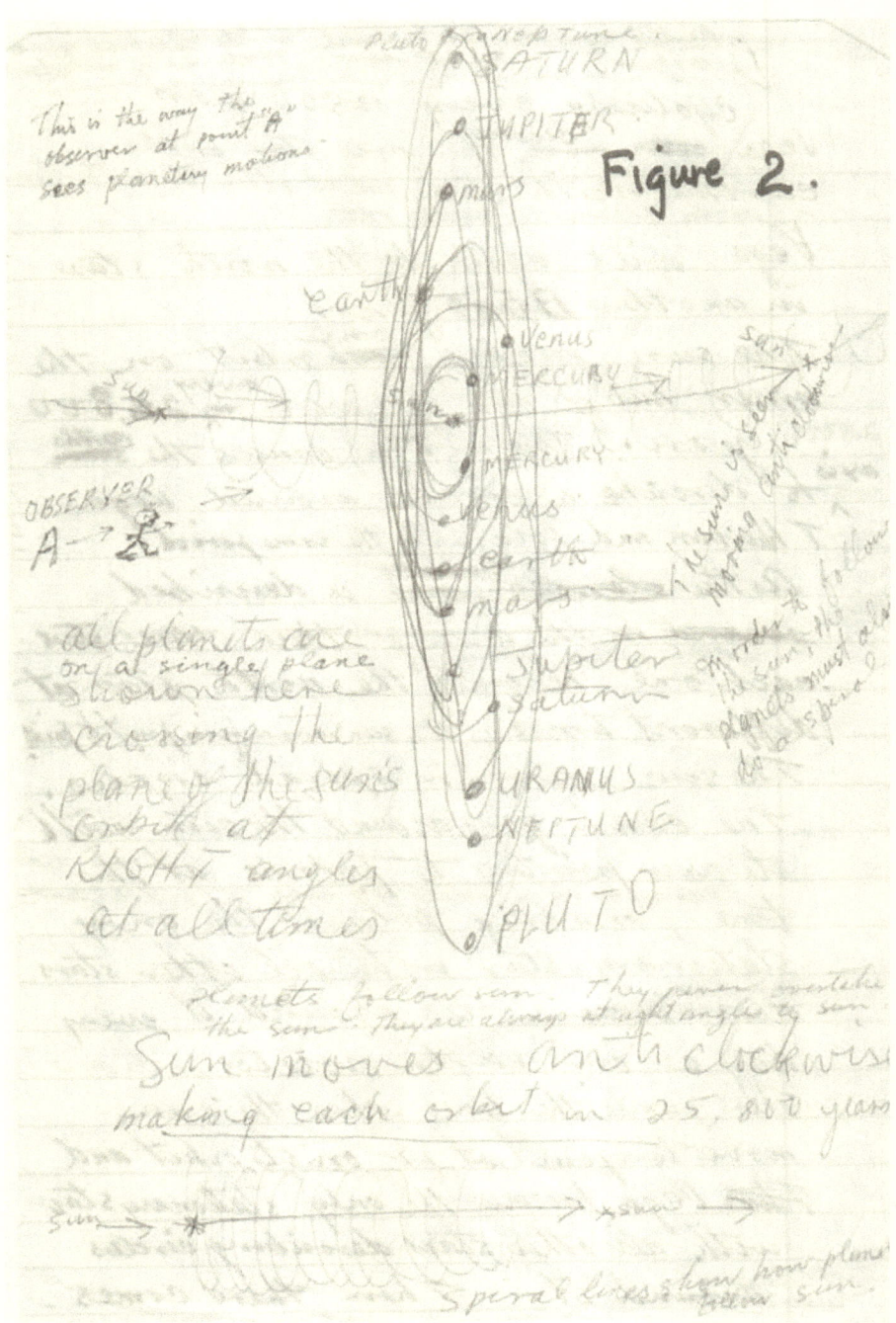

Figure 3: Reversal of Magnetic Poles on Earth

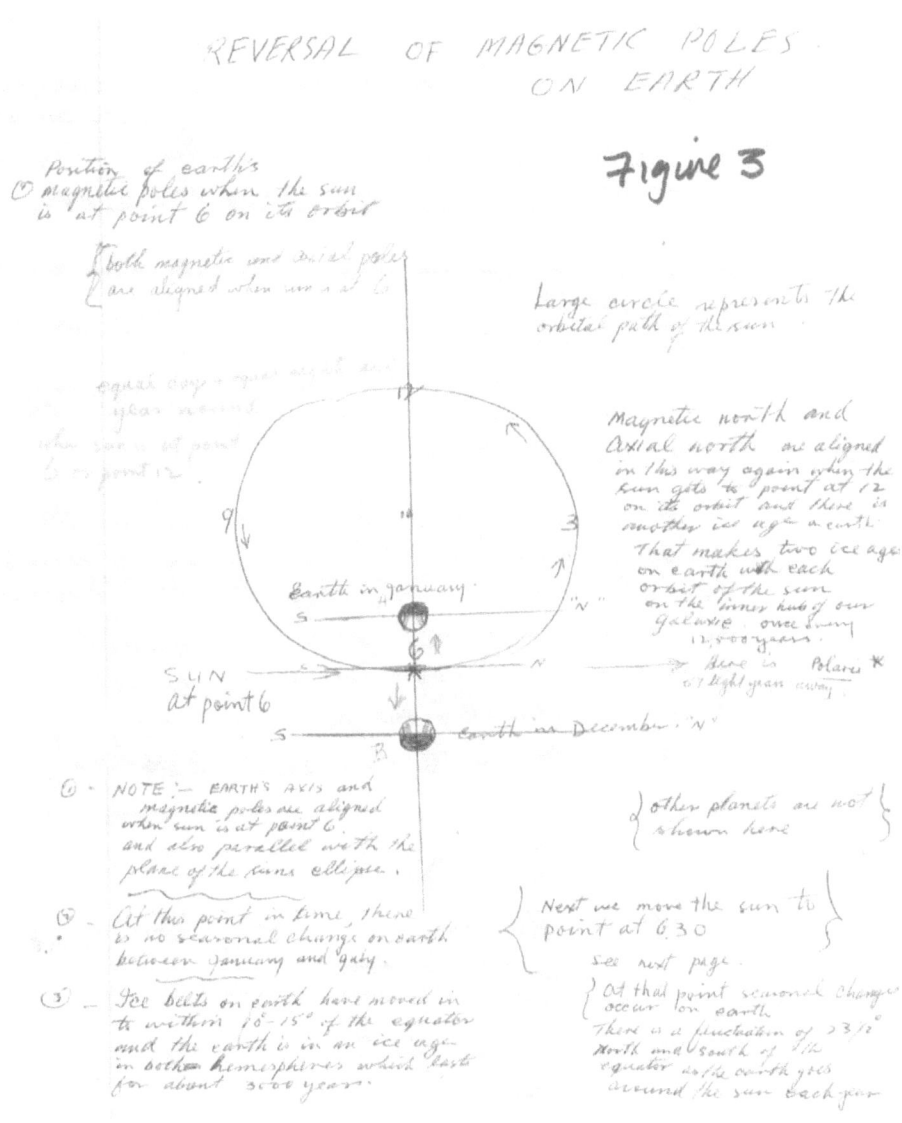

Figure 4: the Mysteries of Climatic Changes on Earth

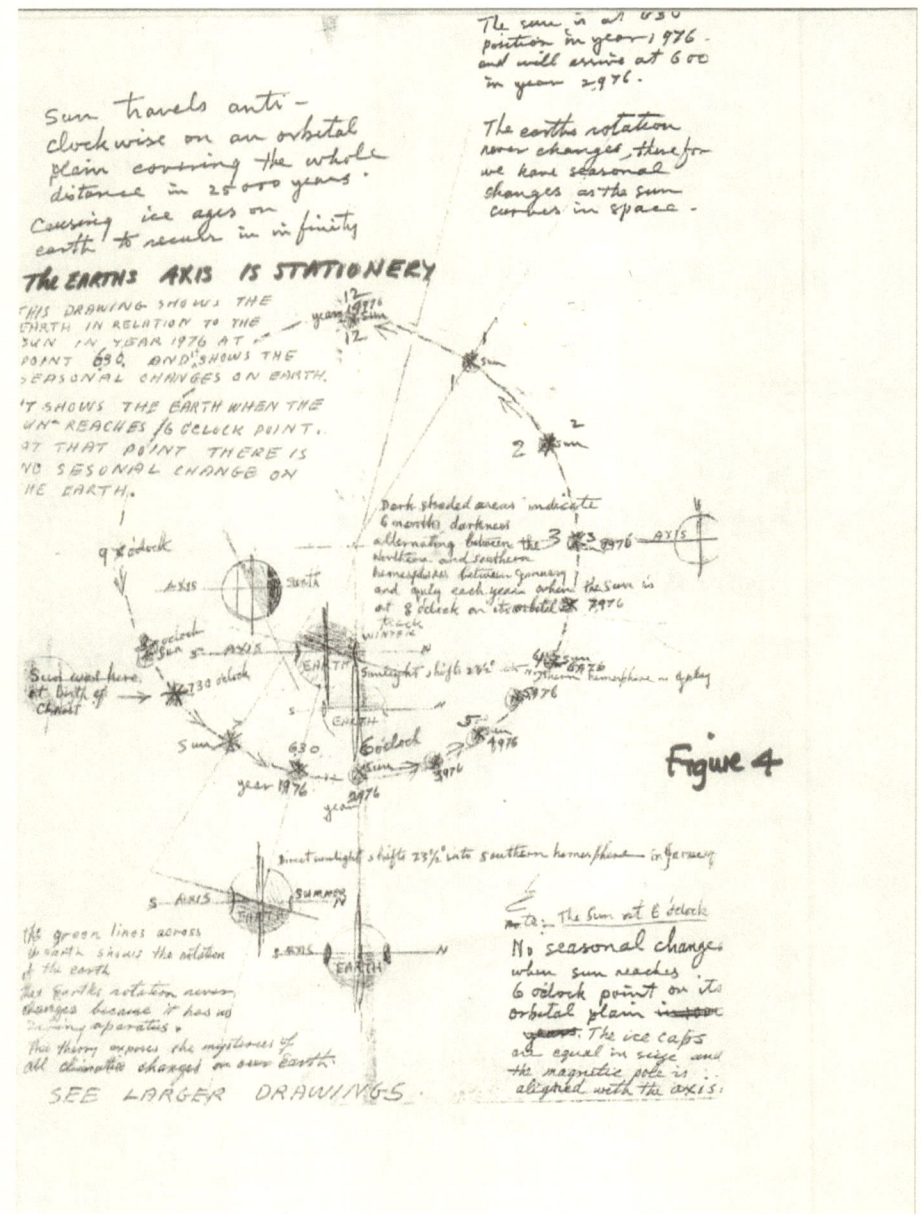

Figure 4

Figure 5: There is no precession nor wobbling of the Earth's axis

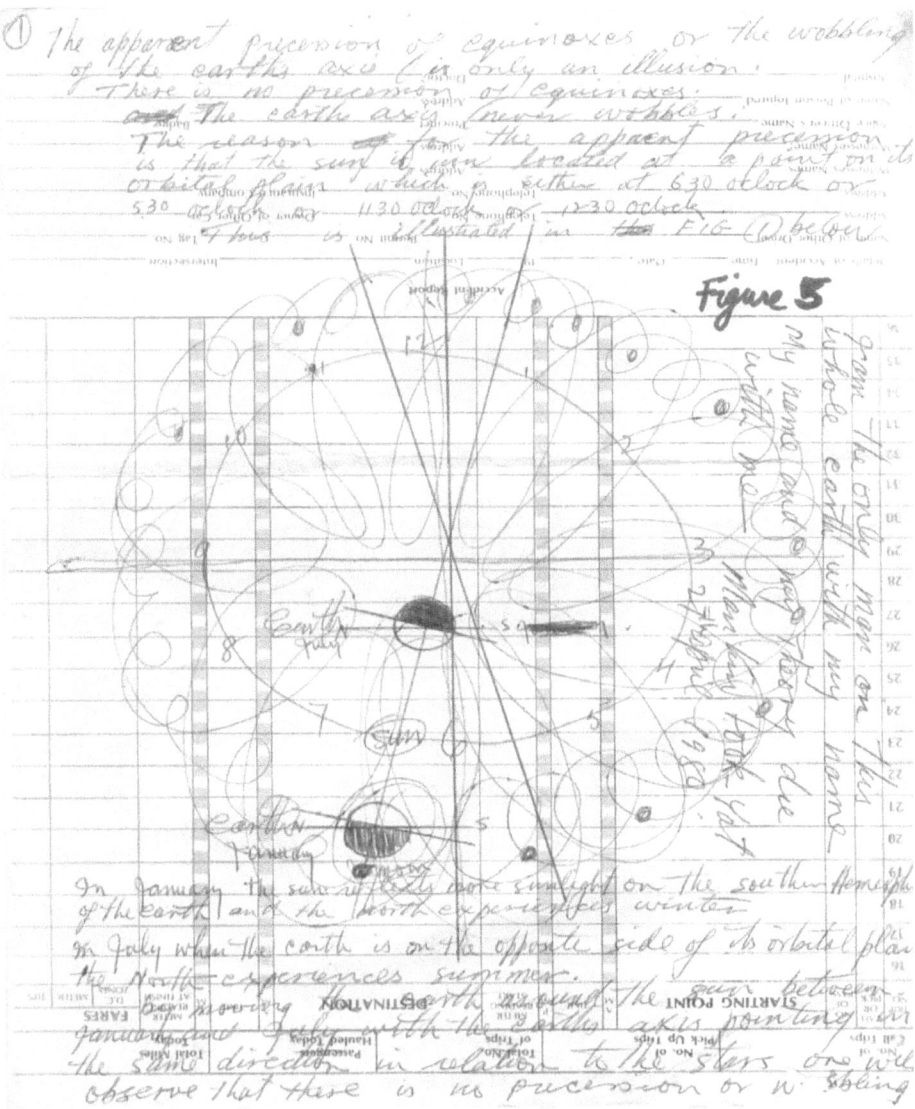

Figure 6: The Earth and Planets never overtake the Sun

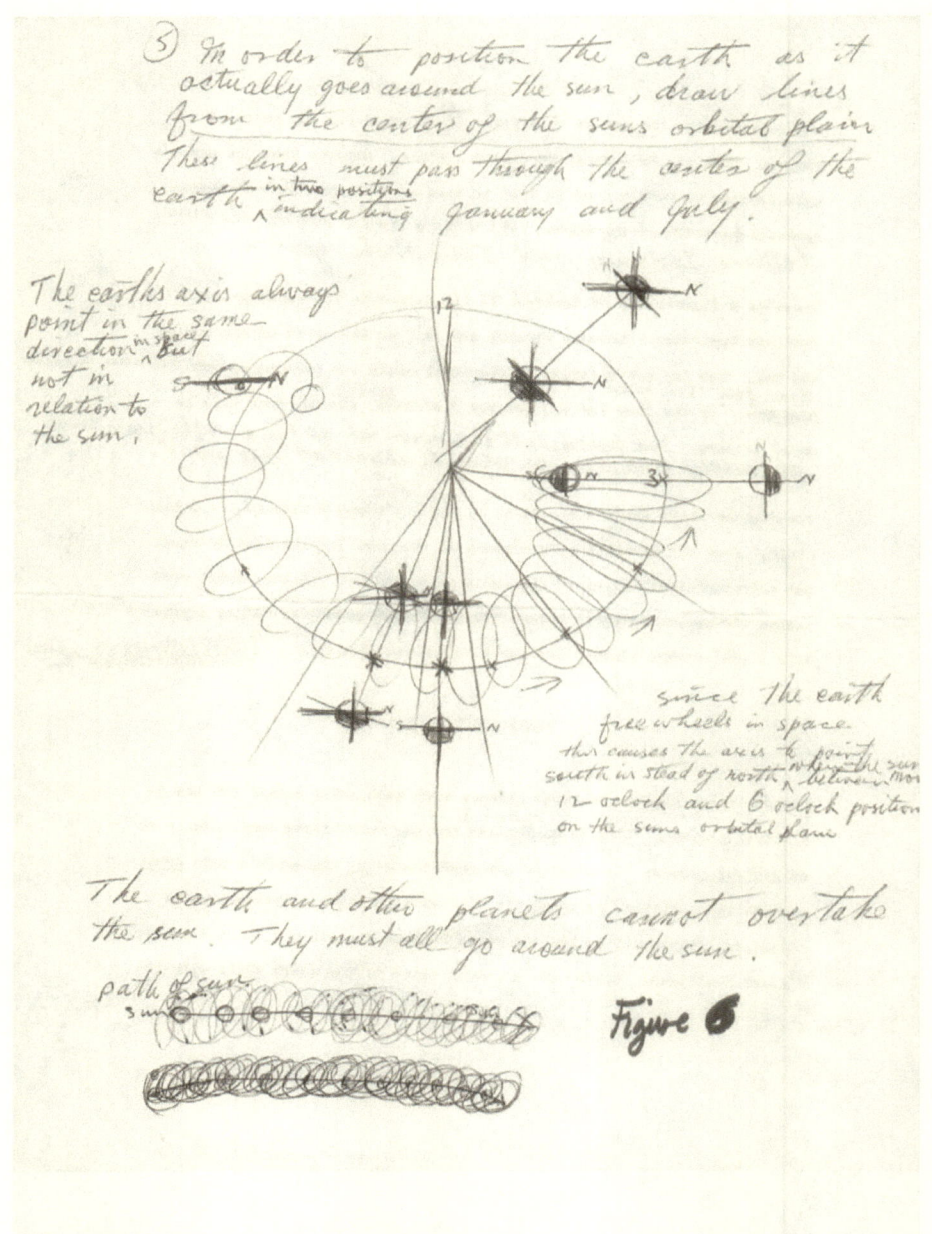

Figure 7: Reversal of Magnetic Poles

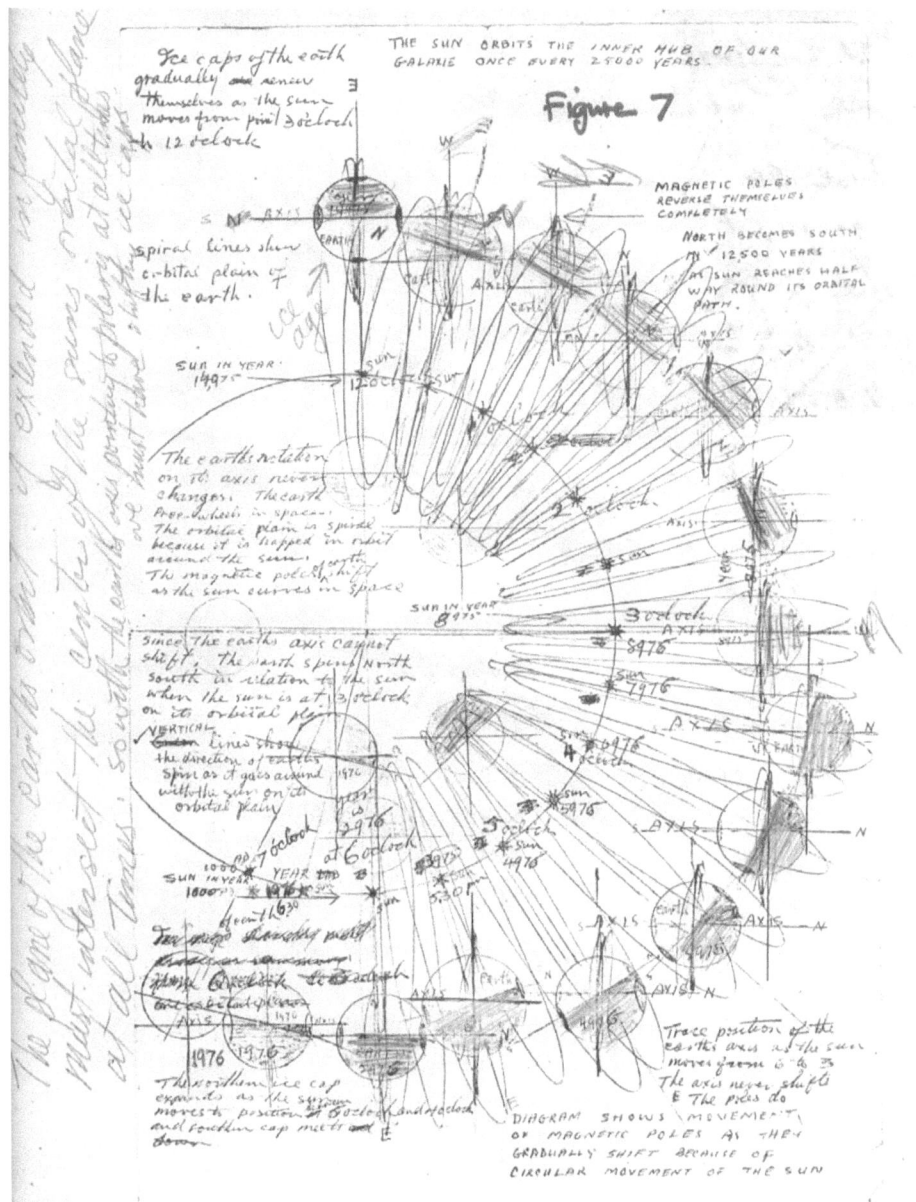

Figure 8: True Planetary Motions

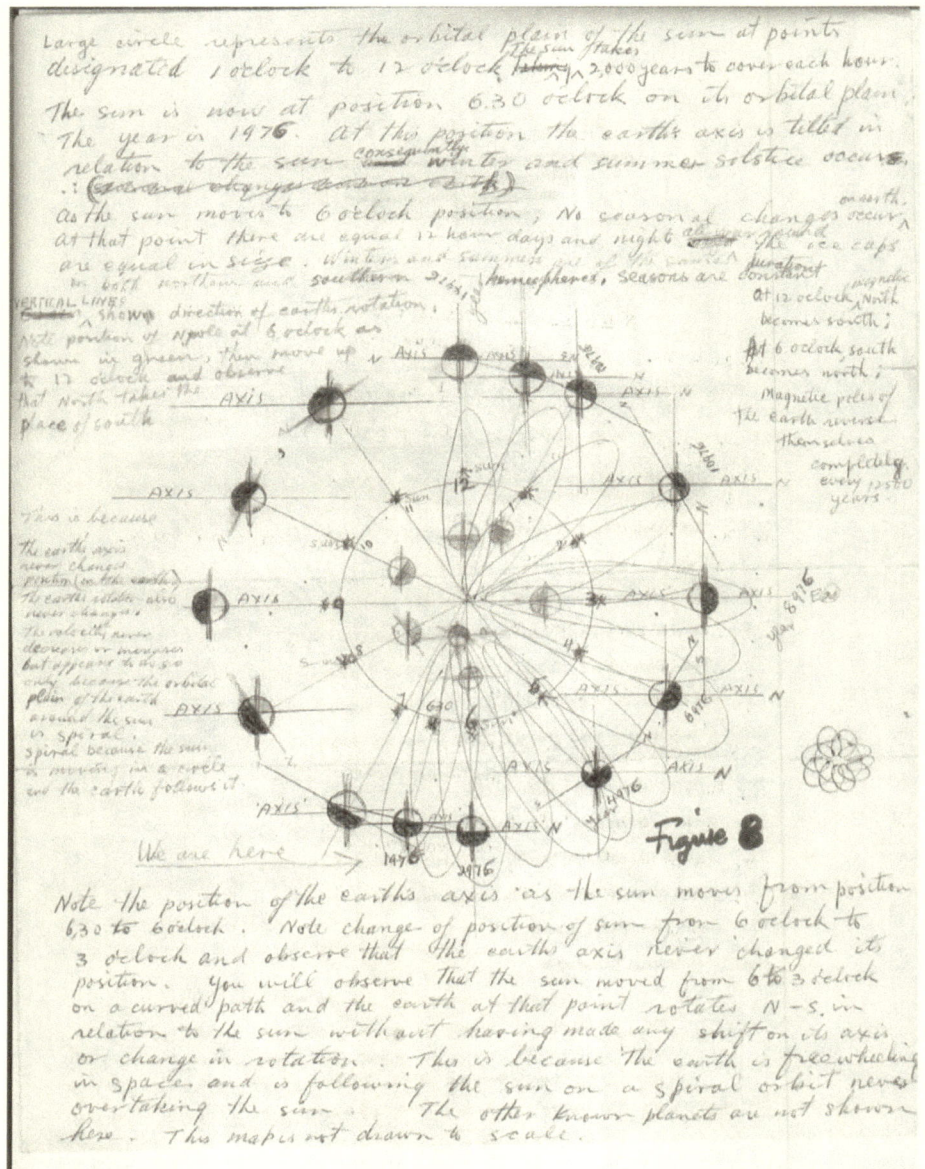

Figure 8

Figure 9: The Ice-ages and Hot Spells in 25,000 year cycles of the Sun

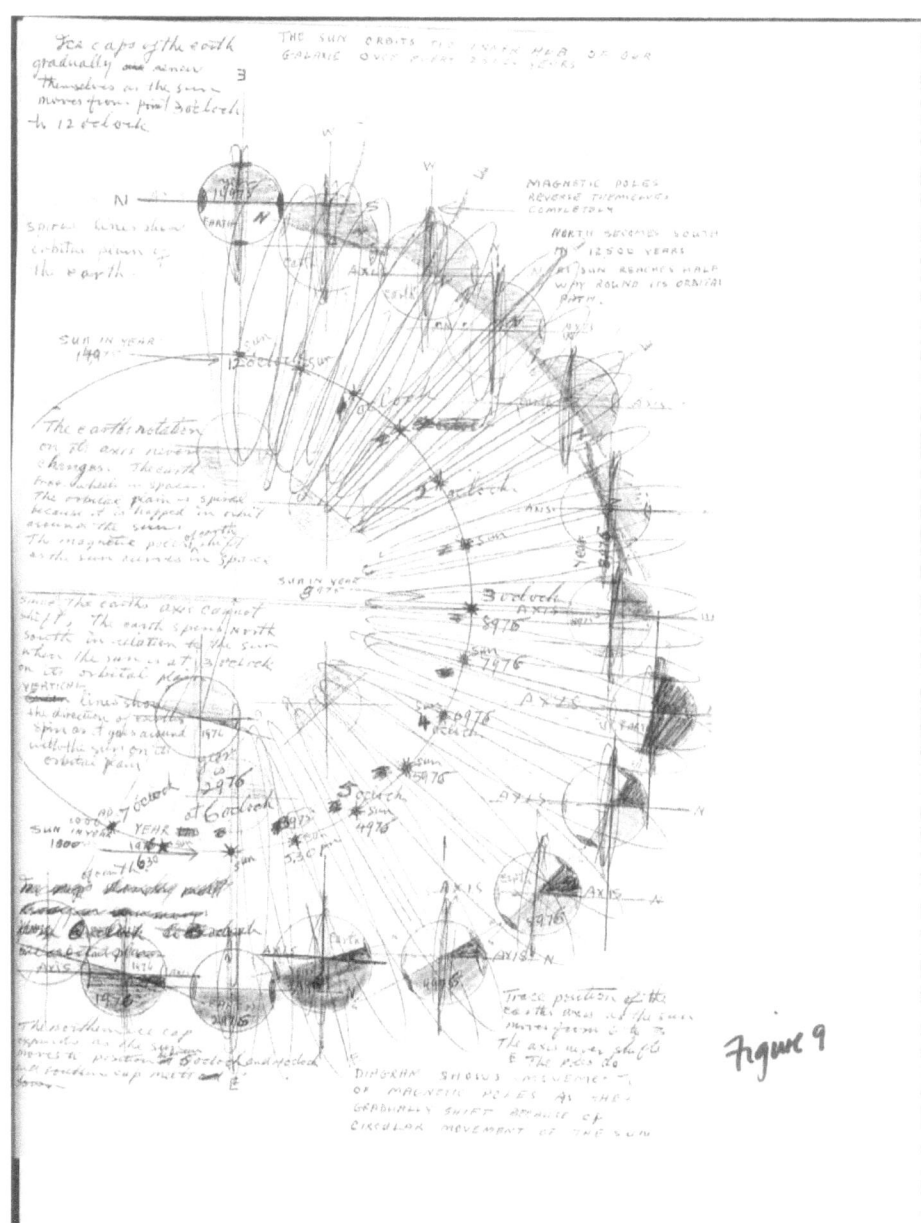

ABOUT THE AUTHOR

Manhin Look-Yat was born in the island of Trinidad and Tobago in 1925 of a native Trinidadian Mother and an immigrant Scottish merchant Father. He was always intrigued as a child by the myriad of life-forms from the minute microcosms of life forms both plant and animal to the vast macrocosm of the universe all around him. His flights of fancy as a youth found him fully engaged in the pursuit of knowledge of such things he was ever so curious about. With a great enthusiasm on the nature of things from geology, anthropology, physics, science in all its strata, to the most deeply rooted fire within, the study of astronomy and the cosmogony of the universe.

By his mid twenties and early thirties he began a series of journals on the various sciences taking great interest in his self-education, and like Author/ Astronomer James Croll before him, not placing emphasis on traditional college education. He valued great thinkers, but challenged the noted learned men of past centuries such as Galileo, Copernicus, and Kepler in his observation of the deficiencies in their projections of planetary motions. He later became engrossed in a particular topic, which even almost 25 years or more ago, placed his theory in an avante-garde category of current proportions in the topic of climatic changes so much a buzz word in today's number one global concerns:

"Global Climatic Changes".

His observation of the universe using first time ever discovered projections from his calculations based on his theory of "TRUE PLANETARY MOTIONS AND RHYTHMIC CLIMATIC CHANGES" is not only brilliantly demonstrated in his documented projections, but will prove to be unfathomably priceless to mankind in the advent of the manifestations and unfolding of his revelations and discovery to Astronomy~ IN THE COMING MILLENNIA AND FOR ALL TIMES.

www.ingramcontent.com/pod-product-compliance
Lightning Source LLC
Chambersburg PA
CBHW022116170526
45157CB00004B/1659